ADVANCED EXPERIMENTAL TECHNIQUES
IN POWDER METALLURGY

PERSPECTIVES IN POWDER METALLURGY
Fundamentals, Methods, and Applications

PERSPECTIVES IN POWDER METALLURGY
Fundamentals, Methods, and Applications

Editors:

Henry H. Hausner
Adjunct Professor
Polytechnic Institute of Brooklyn
Consulting Engineer

Kempton H. Roll
Executive Director
Metal Powder
Industries Federation

Peter K. Johnson
Assistant Director
Metal Powder
Industries Federation

Volume 5

ADVANCED EXPERIMENTAL TECHNIQUES IN POWDER METALLURGY

Based on a Symposium on Advanced Experimental Techniques in Powder Metallurgy sponsored by the Institute of Metals Division, Powder Metallurgy Committee, held at the Spring Meeting of The Metallurgical Society of AIME in Pittsburgh, Pennsylvania, May 1969

Edited by
JOEL S. HIRSCHHORN
Associate Professor of Metallurgical Engineering
University of Wisconsin
Madison, Wisconsin

and

KEMPTON H. ROLL
National Director
American Powder Metallurgy Institute
New York, New York

SPRINGER SCIENCE+BUSINESS MEDIA, LLC 1970

The editors gratefully acknowledge permission to reprint the following article, which appears as the last paper in this volume: A Literature Review of Mercury Porosimetry and a Discussion of Possible Sources of Errors in the Method, by H. M. Rootare, *Aminco Laboratory News,* Vol. 24, No. 3, 1968

Library of Congress Catalog Card Number 78-127938

ISBN 978-1-4615-8983-9   ISBN 978-1-4615-8981-5 (ebook)
DOI 10.1007/978-1-4615-8981-5

United Kingdom edition published by Plenum Press, London
A Division of Plenum Publishing Company, Ltd.
Donington House, 30 Norfolk Street, London, W.C.2, England

PREFACE

The increasing use of powder metallurgy techniques to make an almost infinite variety of materials and products places greater emphasis on utilization of sophisticated experimental techniques. Usually research and development efforts initiate the use of newly developed equipment and analytical procedures. Indeed, the contents of this book are strongly linked to research endeavors, in both the academic and industrials worlds.

However, this volume can serve a much needed function in industrial applied powder metallurgy. Although many researchers will find the contents of great value, the technical personnel more involved with production, quality control, customer services and product design now have at their disposal a means to learn about the potential uses of several very important techniques. With today's "knowledge explosion" the present set of papers greatly facilitates the comprehension and adoption of new procedures.

If powder metallurgy is to continue its rapid rate of growth in virtually all segments of industry, then the transition of modern equipment and procedures from tools of research and development laboratories to everyday plant operations and applications must be hastened.

The editors hope that this volume aids in this process, as well as assisting students and researchers by providing a ready source of up-to-date useful information.

Joel S. Hirschhorn
Kempton H. Roll

CONTENTS

SCANNING ELECTRON MICROSCOPY IN POWDER METALLURGY

O. Johari

IIT Research Institute
10 West 35th Street
Chicago, Illinois 60616

A. INTRODUCTION

The number of scanning electron microscopes and their applications are increasing very rapidly. The easily interpretable three-dimensional images produced by the SEM have been drawing considerable attention in all fields of science and technology.[1-3] In a recent paper, Johari and Bhattacharyya [4] have described the application of scanning electron microscopy in the characterization of powders produced by a variety of processes. The SEM has wide applications in all fields of powder metallurgy where a knowledge of surface characteristics, such as size, shape, porosity, contamination, and topography, is essential.

The use of optical microscopes in examining rough surfaces is restricted by their small depth of focus. Transmission electron microscope replica techniques suffer from numerous disadvantages; these have been reviewed in a comparison paper.[5] Here the working of the scanning electron microscope is explained with suitable examples.

B. How a Scanning Electron Microscope Works

Figure 1[6] shows the schematic diagram of
fundamental components of the SEM. The beam of
electrons produced at the electron gun is accel-
erated to a potential of 10-25 kv and is focused
by a system of lenses on the specimen surface
where a finely focused image of the electron
source is formed. The SEM image can be produced
by any signal generated by this primary beam as
it scans over the specimen surface. The cathode
ray tube on which the image of the specimen is
formed is simultaneously scanned with the electron
beam on the specimen surface. The primary electrons can produce low energy secondary electrons,
back-scattered electrons, induced currents,
photons, x-rays, absorbed electrons, and some heat
effects. Secondary electrons are commonly used
to form topographical images of the surfaces.

Each primary electron produces a large number of low energy secondary electrons which are
diverted to a collector system by applying a
10-12.5 kv positive potential. The signal received at the detector is amplified and is used
to modulate the image on the CRT. As the primary beam scans over the surface, the image on
the CRT is built up of variations in the number
of secondary electrons reaching the collector
from each point hit by the primary beam on the
specimen surface. At first glance the picture
resembles a dark-field optical micrograph of a
surface illuminated from one direction (electron-
collector direction), but subsequent observation
reveals the following important observations:

   (1) The large positive potential at the
   detector can attract secondary electrons
   by bending them from surfaces otherwise
   not in its direct view. This means that
   re-entrant holes or crevices along the
   axis of the elctron beam can also be examined. Of course, we cannot see those
   hidden surfaces where primary electrons
   do not reach, such as those at undercuts
   or around the point where a sphere touches

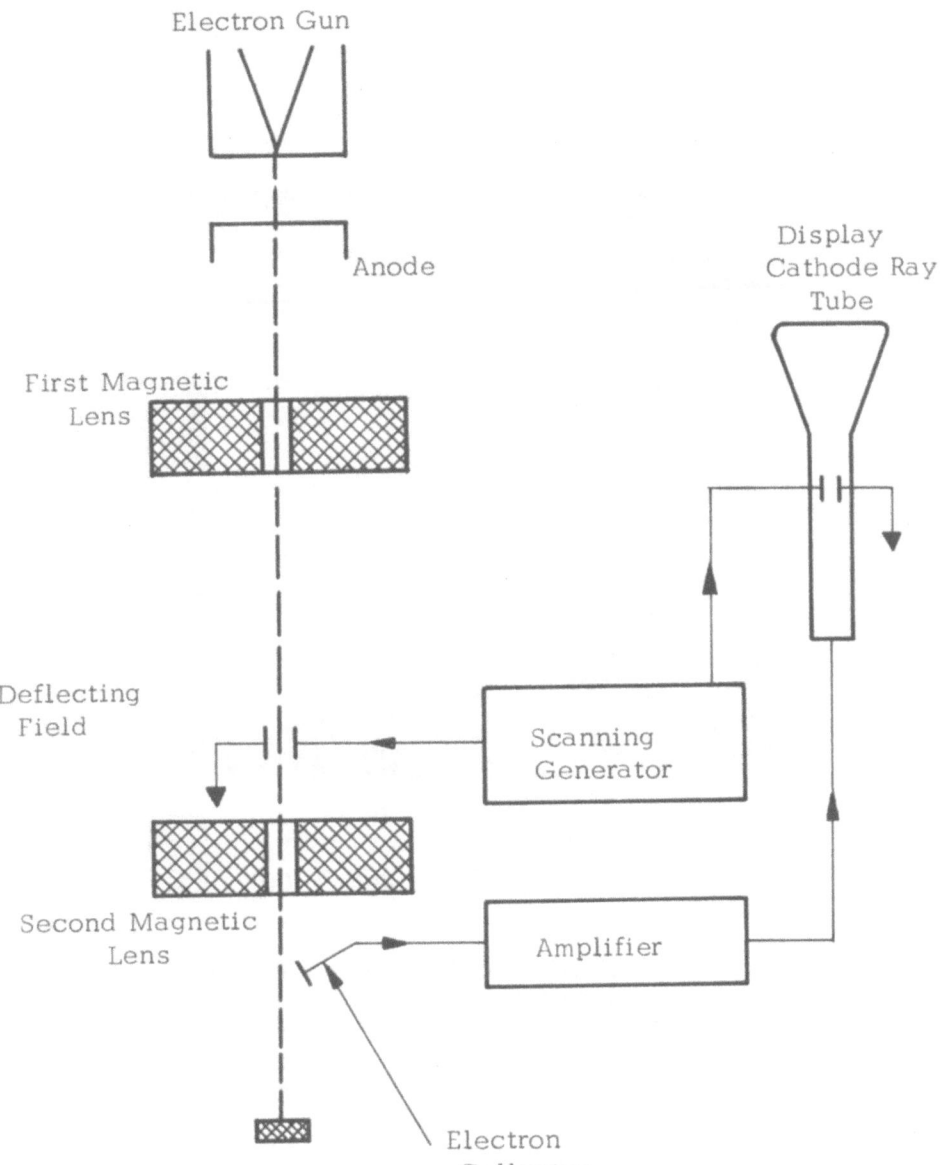

Fig. 1

PRINCIPLE OF THE SCANNING ELECTRON MICROSCOPE

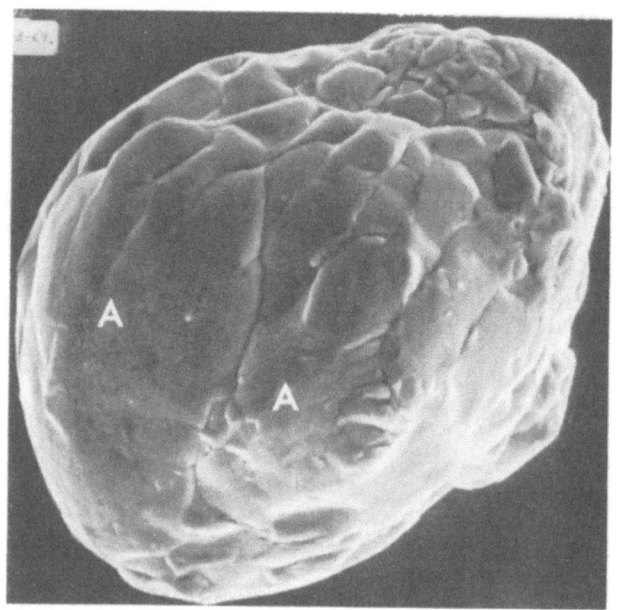

Fig. 2
Tilt 0°, Rot. 0°, Mag. 1000X

Scanning electron micrograph of an
atomized Inconel 713C powder. (The
direction of detector is along the
right-hand side of the photograph,
the area which appears excessively
bright.)

a surface   (Figures 2,3, and 4).

(2)   Sharp points, edges, and similar geometrical shapes which can emit electrons from many directions always appear excessively bright and, thus, require special care in photographing and interpreting the results.

(3)   Depending on the geometry of the surface, some back-scattered electrons will also reach the collector, and this may cause shadow effects similar to those produced in dark-field optical microscopy.

(4)   The number of secondary electrons reaching the collector depends on the nature of the surface--i.e., denser materials having more electrons will emit more electrons compared to lighter materials.   Thus, contamination of organic films, etc., is easily recognized as dark, structureless regions (Figure 5).   Also, the back-scattered electrons are appreciably affected by emission effects due to atomic number differences, and on a polished surface bright features can be identified as being from high-average atomic number regions.

(5)   The charge given to the specimen by the impinging primary beam must always be carried away.   This is no problem for conductive materials, but for nonconductors, a thin coating of conductive material should be applied to leak the charge away to earth. Even if the material is conductive, the presence of a local non-conductive coating may cause locally charged-up areas; these areas appear bright because the more charged-up areas will emit more secondary electrons (Figure 6).

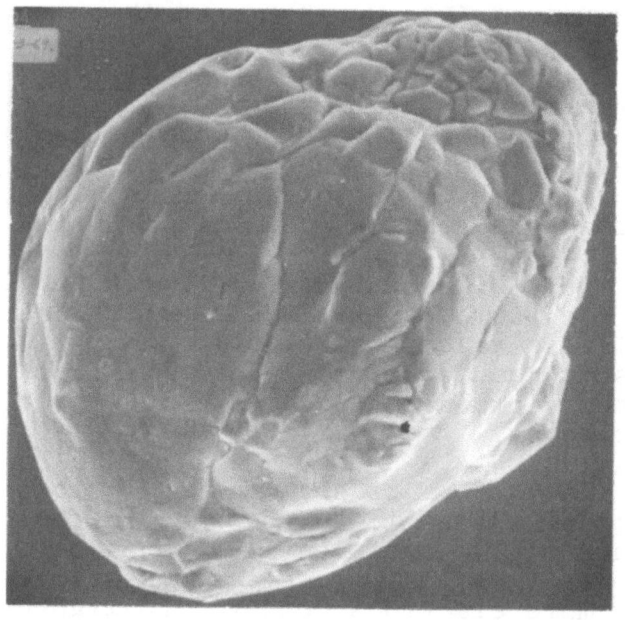

Fig. 3

Tilt 6°, Rot. 0°, Mag. 1000X

This photograph, along with Fig. 2 (LHS and RHS, respectively), constitutes a stereo-pair, and shows that the particle is a deformed sphere. The surface being examined is a distorted sphere, whose highest point in this view is at A points, such as along the left-hand side of the photograph, and should be in shadow so far as the position of the collector is concerned. They are seen because, as explained in text, the positive potential at the detector can attract the low-energy secondary electrons to it from behind surfaces. The stereo view will also show that no points on the bottom half of this sphere are revealed because they are not hit by the primary beam, and hence cannot produce any signal to reach the collector.

Fig. 4   These eight views are of the same par-
          ticle in Figures 2 and 3, but viewed at
          different angles. Seeing them all to-
          gether allows one to view most of the
          surface of the particle. This freedom
          in SEM is one of its many unique advant-
          ages; however, one cannot view the re-
          gion along which this particle rests on
          the support surface.

          A three-dimensional model of the part-
          icle can be constructed and verified
          using these many views, along with ap-
          propriate stereo-pairs.

Fig. 5   Figure 5a is a low magnification photo-
          graph showing dendritic structure in
          atomized Astroloy (U-700) powder.
          Figures 5b, c, and d show high magni-
          fication views, all at the same mag-
          nifications, of areas P, Q, and R, re-
          spectively, in (a).

          The contrast at the dark region in b is
          attributed to an organic film. SEM can
          thus be used to show presence of film
          or second-phase contaminants on powder
          surfaces. The dendrite spacing, etc.,
          on the sphere surface is a function of
          the alloy composition and its cooling
          rate.

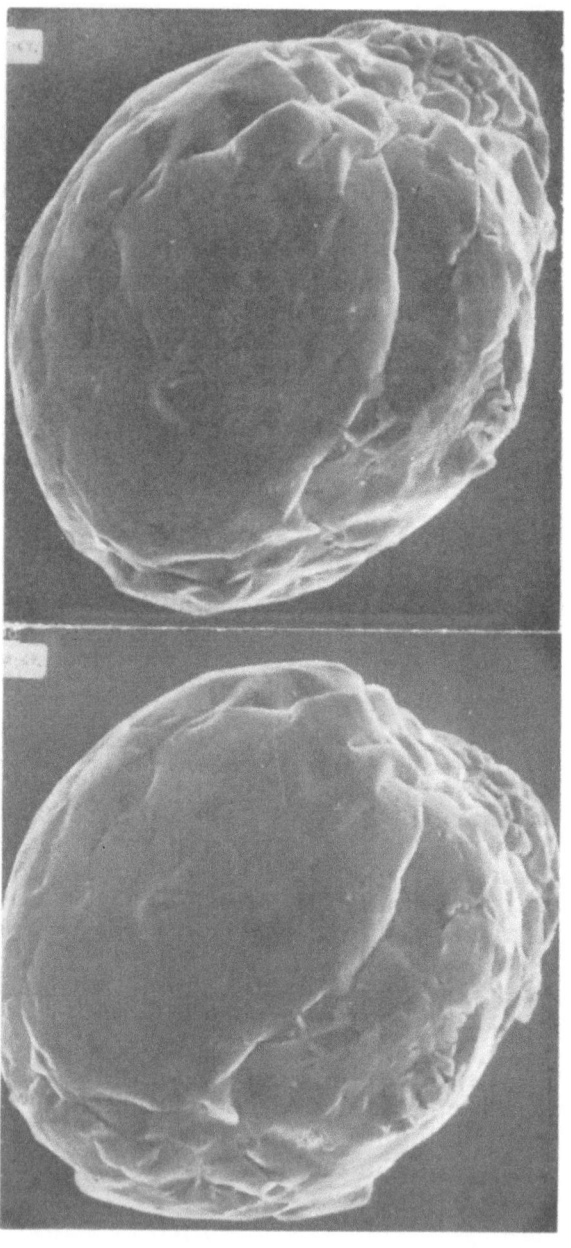

Fig. 4(a)   Top, Tilt 45°, Rot.   0°, Mag. 1000X

Fig. 4(b) Bottom, Tilt 45°, Rot. +30°, Mag. 1000X

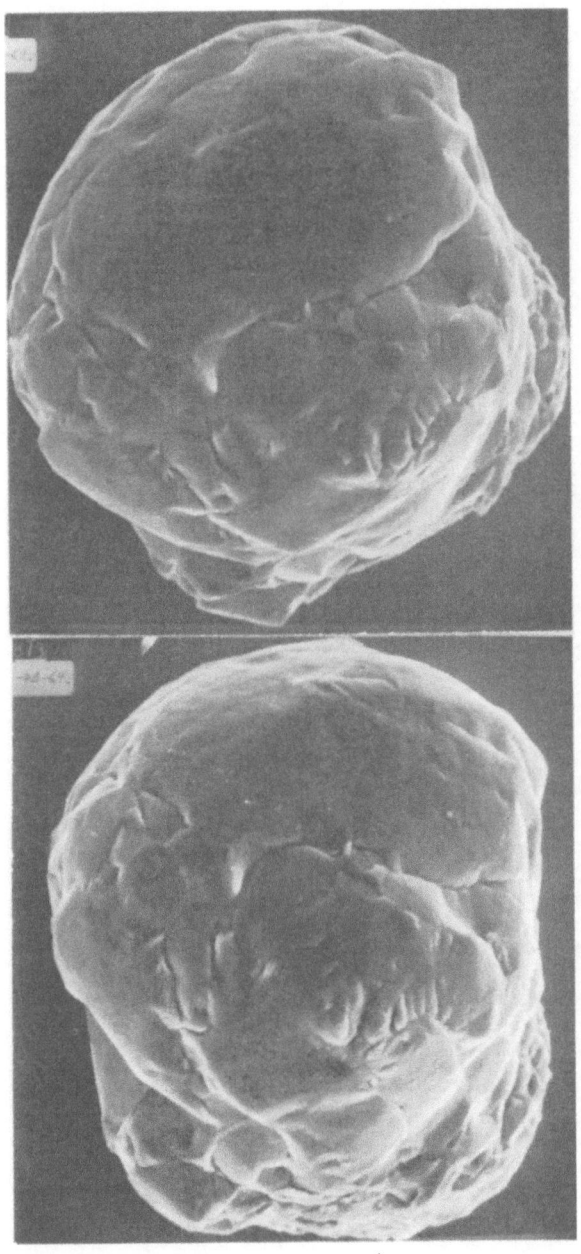

Fig. 4(c)   Top, Tilt 45°, Rot. +60°, Mag. 1000X

Fig. 4(d) Bottom, Tilt 45°, Rot. +90°, Mag. 1000X

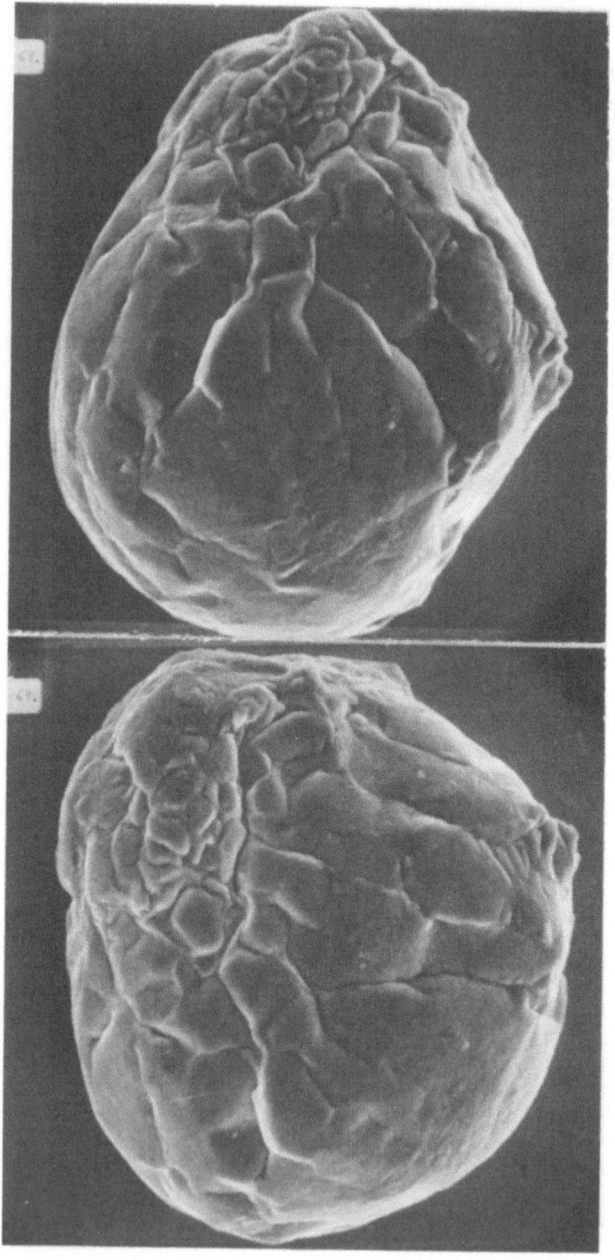

Fig. 4(e)   Top, Tilt 45°, Rot. -45°, Mag. 1000X

Fig. 4(f) Bottom, Tilt 45°, Rot. -90°, Mag. 1000X

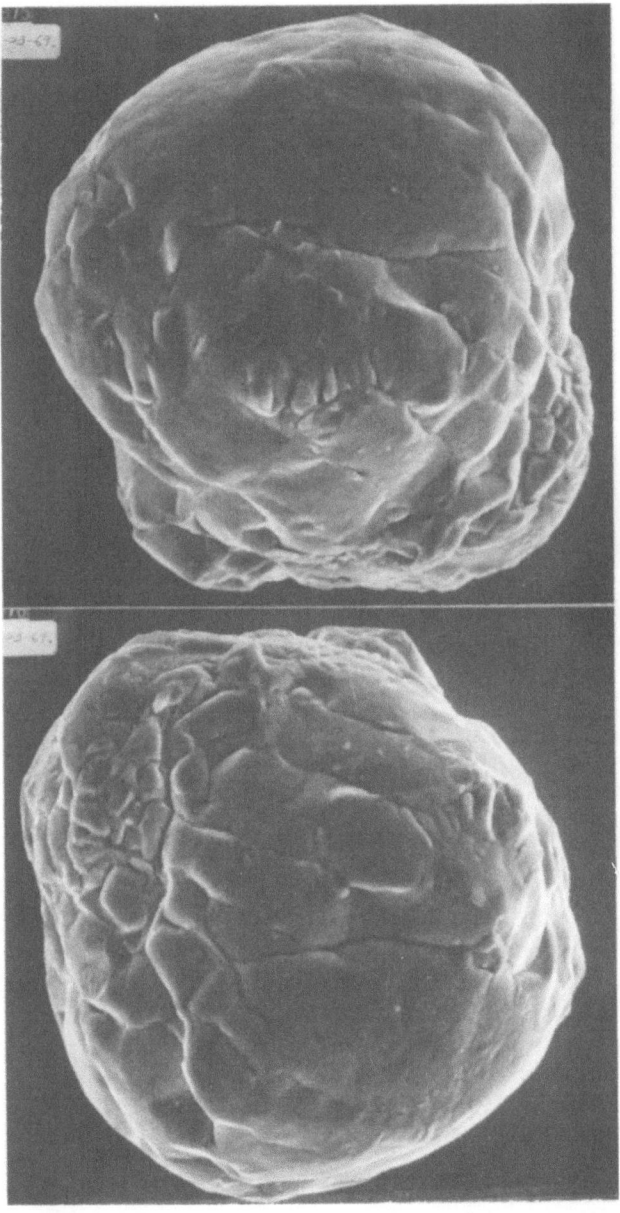

Fig. 4(g)   Top, Tilt 25°, Rot. +90°, Mag. 1000X

Fig. 4(h) Bottom, Tilt 25°, Rot. -90°, Mag. 1000X

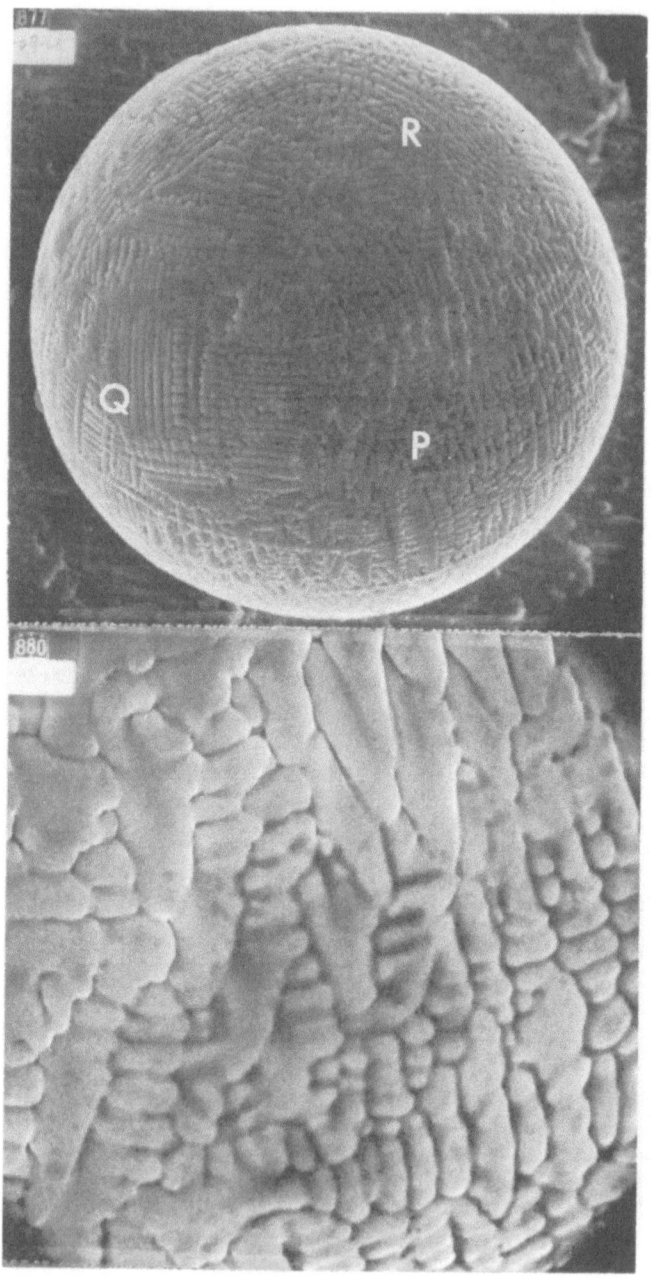

Fig. 5(a)   Top, Mag. 1150X

Fig. 5(b) Bottom, Mag. 4000X

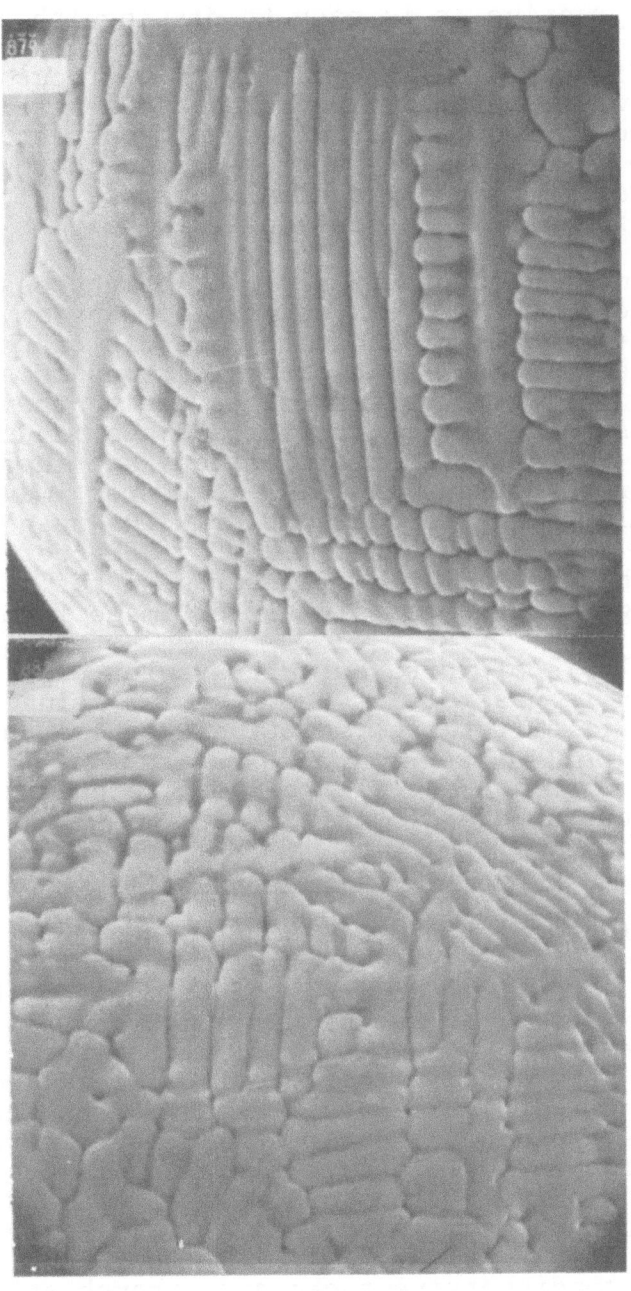

Fig. 5(c)   Top, Mag. 4000X
Fig. 5(d) Bottom, Mag. 4000X

In practice it is easy to distinguish contrast caused by charging-up or atomic number effects (Figure 6). The great depth of focus of the SEM allows topographical contrast to be easily separated from the other causes of contrast. More and more, use is being made of stereo photography to separate true topographical contrast. This application will be discussed separately below.

As mentioned previously, the brightness of the image on the CRT can be modulated by any of the other signals produced as a result of the primary beam. Back-scattered electrons on smooth surfaces will give better images due to their greater sensitivity to surface and atomic number effects. Unless electrical effects are involved, the absorbed current and back-scattered images are complementary to one another. The resolution in these as well as other modes of SEM operation is poor because a large diameter beam has to be used to get sufficient signal for image formation.

Analysis of infrared, ultraviolet, and light radiation is possible if the material being examined will emit these radiations. These analyses--as well as auger, back-scattered, and absorbed electrons--can also be used to chemically analyze the nature of surface.

C. Possibilities with the SEM

Besides topographical work, SEM offers many other possibilities of interest to powder metallurgists. Chemical analysis of the surface by the above mentioned techniques and x-rays is possible. X-rays produced as a result of the primary electron beam contain characteristic peaks corresponding to the elements present. With the appropriate placement of crystal detectors (similar to those in the microprobe) or nondispersive Li drifted silicon detectors, elements and their distribution can be determined.(7)

The nondispersive technique, though still in a continuous state of development, is very rapid and offers considerable promise as a universal

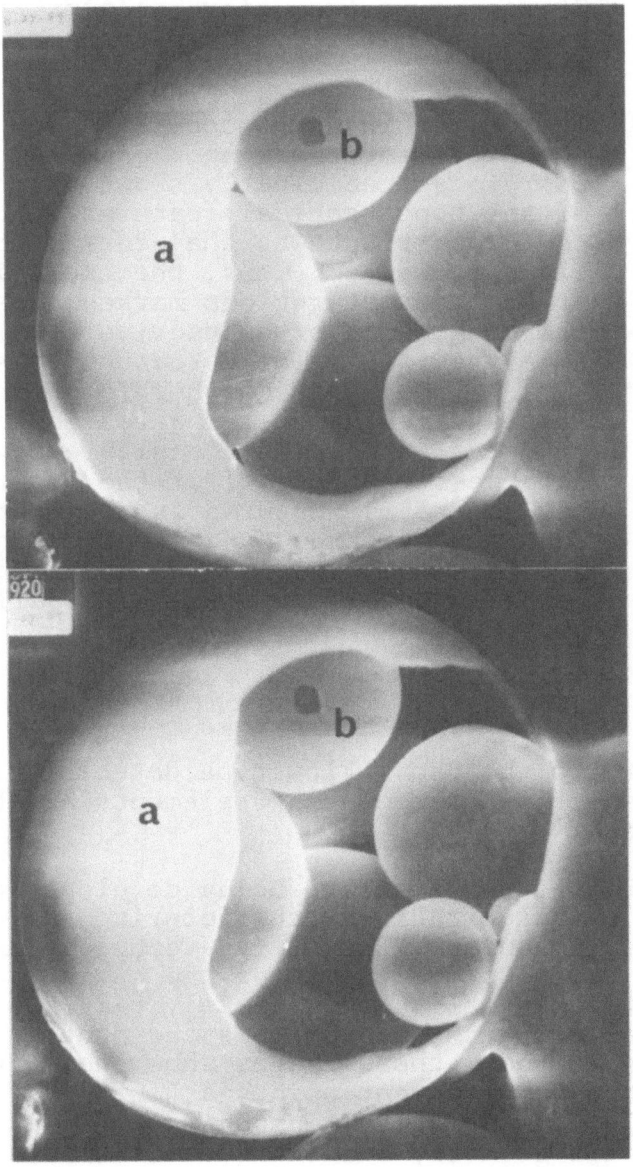

Fig. 6(a) Top 0° Tilt, Mag. 1000X

Fig. 6(b) Bottom 6° Tilt, Mag. 1000X

Fig. 6   Stereo-pairs from microballoons. (Mi
           cron sized hollow sphere of phenolic
           resin).

           These organic compounds were gold coat-
           ed to prevent charging. Viewing of the
           pair in stereo clearly shows absence
           of any topographical features which
           could contribute to the contrast of re-
           gions marked a. Thus, the excessively
           bright area at regions marked a is due
           to the absence of conductive coating in
           local regions. The structure consists
           of a hollow, broken sphere containing
           at least seven other spheres inside it.
           From the three-dimensional viewing, the
           thickness of the wall of the outer bro-
           ken sphere is measured as one micron.
           Most spheres are connected to each oth-
           er. The dark spot at b is a hole; this
           sphere probably contains more balloons.

accessory to the SEM. Flow type detectors for
analysis of low atomic number elements are being
developed. (8)

     Stereo techniques are being developed with
the SEM to quantitatively characterize the topographical features. The stereo-technique neces-
sitates two recorded images of the surface from
two different angles approximately 6-10° apart.
When analyzed with stereoscopes and photogram-
metry techniques, these images show the heights
and depths of different points as in topographical
maps. This information, in conjunction with three
dimensional geometry can be used to calculate many
other parameters such as volume and surface area.
In sintering experiments, this will provide for
a very accurate measurement of neck radius. In
a paper of this nature, it is not possible to
present results in three dimensions. Examples
of stereo-pairs from powders are presented in

Figures 2, 3, and 6. Instead of tearing the page
and viewing through a stereo-viewer, the reader
may find it very illustrative to have stereo-
slides prepared and viewed in a stereo-slide viewer or projector.

The value of stereo-photographs as an aid in
interpretation of exact topographical detail of
the surfaces is being realized more and more by
the SEM users. Many laboratories are working on
the placement of an additional set of deflection
coils, so that tilting of samples to get the
stereo effect can be achieved by flicking a
switch. This method is similar to the production
of bright and dark-field photographs in the transmission electron microscope. In TEM, a set of
deflection coils with azimuth and magnitude control can be properly placed so as to deflect the
primary beam to impinge on the sample along one
of the deflected beams.

Actual three-dimensional models of surfaces
can be constructed at any desired magnification
by using stereo pairs and the tilt and rotation
capability of the SEM (e.g., from various views
presented of same powder in Figures 2,3, and 4).

The value of such models in the clear understanding of various phenomena under study does
not need elaboration. Examples of such work in
dentistry have been discussed by Boyde, [9] and
with increasing uses of SEM, such examples should
become increasingly available in powder metallurgy.

Thus, the SEM provides the researcher with a
tool for complete three-dimensional characterization of a surface at all available magnifications,
including actual models of surface, contour maps,
and profile diagrams, and height, depth, volume,
and surface area measurements.

Another attractive feature of the SEM is the
television accessory for continuous scanning of
the area being imaged. (10) This accessory allows,
at a slight sacrifice of resolution, video-taping
of dynamic events as they occur. Thus, dynamic
experiments in the sintering of powders and the
formation of powders by chemical reaction techniques becomes increasingly feasible. It is safe
to predict that knowledge gained about true surfaces through the SEM and its various capabilities
should confirm or radically alter our concepts of
powder metallurgy phenomena.

D. Examples Illustrating Features of the SEM

Figures 2 to 6 illustrate two of the basic
features of the SEM well--i.e., a large depth
of focus and direct examination. The technique
is exceptionally rapid; the total time required
to obtain the photographs in Figures 2, 3 and 4
was only 2 hours. The ability to view powders at
different angles is easily illustrated by example
of a cube. One may examine five of the six cube
faces. This SEM feature is shown in Figure 7
for carbonyl nickel particles forming single
cubical crystals.

SEM examination of areas at varying magnifications is illustrated in Figure 8. These micrographs, from a sample of TiO2, also illustrate
the need to actually see the powders in order to

Fig. 7   Figures 7a and b are stereo-pairs, as
           are Figures 7c and d. By changing ro-
           tation and tilt, details of other faces
           of the cube are revealed. Since a
           sample can be tilted and rotated much
           more than illustrated here, five of the
           six faces of the cube can be readily
           examined.

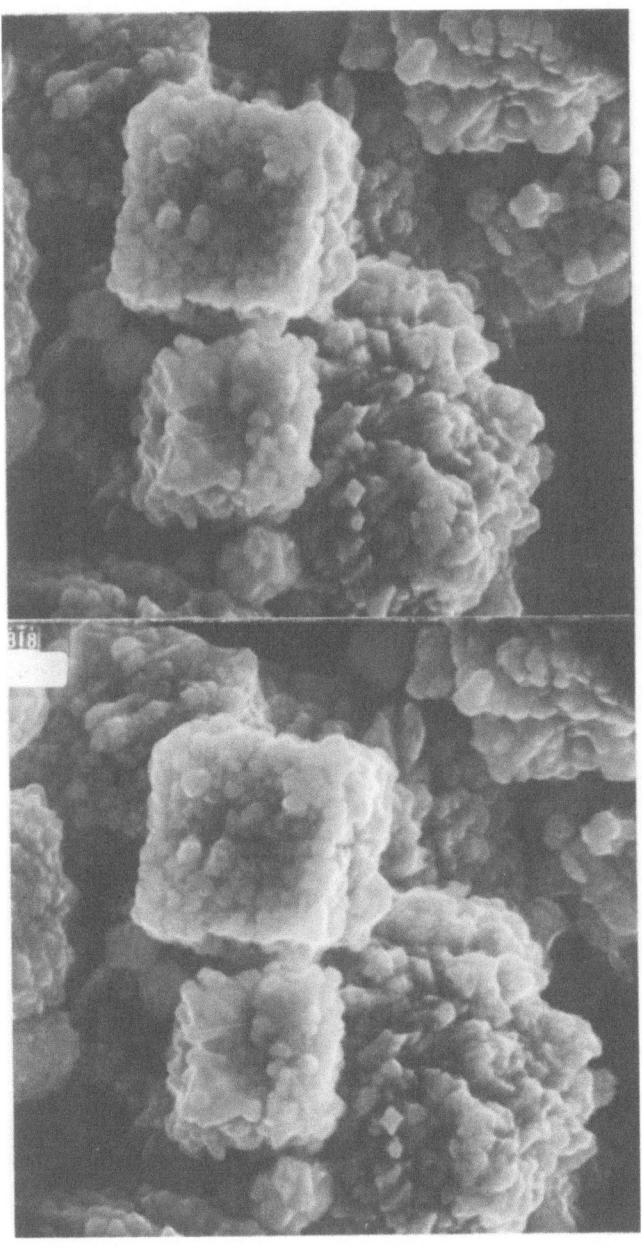

Fig. 7(a)   Top, Tilt 19°, Rot. -90°, Mag. 10,000X

Fig. 7(b) Bottom, Tilt 25°, Rot. -90°, Mag. 10,000X

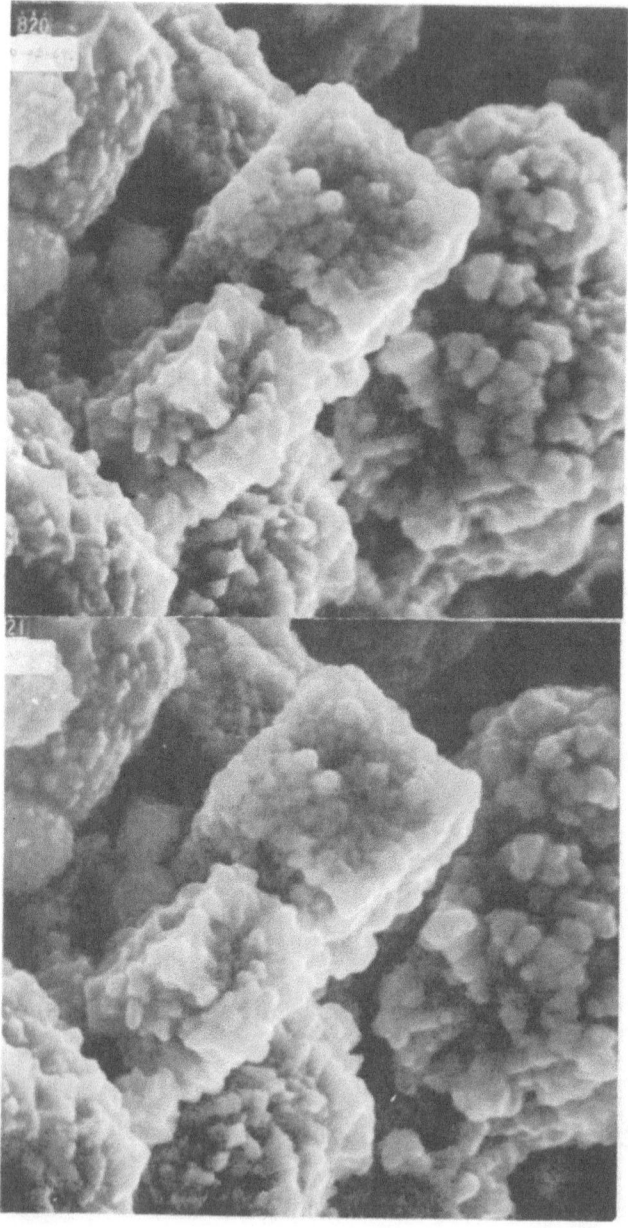

Fig. 7(c)   Top, Tilt 45°, Rot. -45°, Mag. 10,000X
Fig. 7(d) Bottom, Tilt 39°, Rot. -45°, Mag. 10,000X

support the indirect techniques used to determine
the size and shape of the particles. Development
of proper sampling and distribution techniques,
along with computer processes to quantitatively
calculate size and shape parameters from SEM
images, are now being attempted to increase the
usefulness of the SEM.

E. Conclusions

    Rapid developments are taking place in the
SEM instrumentations, accessories, techniques,
and applications so that the "pretty" pictures obtained directly from examination of powders and
powder related phenomena can be used to determine
many quantitative and qualitative parameters.
Many features of this instrument, along with ex-
citing possibilities offered by the stereo techniques, analytical results, and dynamic experiments make this instrument a must in the study of
almost all phenomena related to the study of
powders.

Fig. 8   This represents four views of TiO_2
            powders. The areas examined in higher
            magnification photographs are marked
            on preceding low magnification photo-
            graphs. The micrographs show the need
            to actually examine the powders before
            commenting on their size and shape
            parameter, as well as agglomeration
            tendencies.

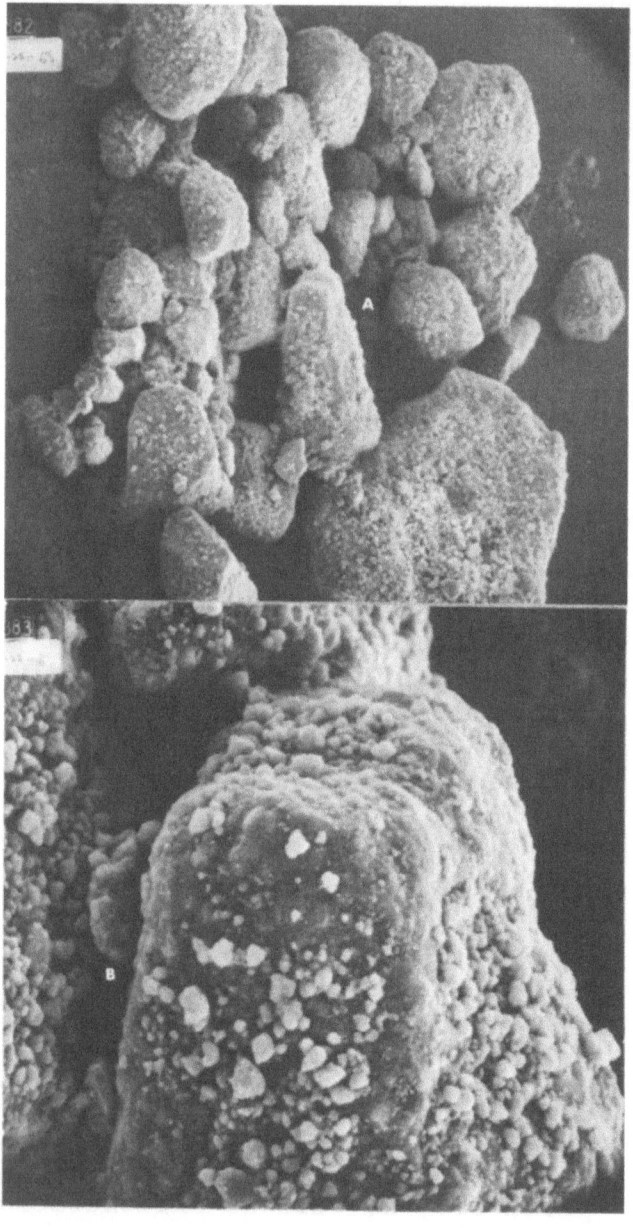

Fig. 8(a)　　Top, Mag. 100X
Fig. 8(b) Bottom, Mag. 700X

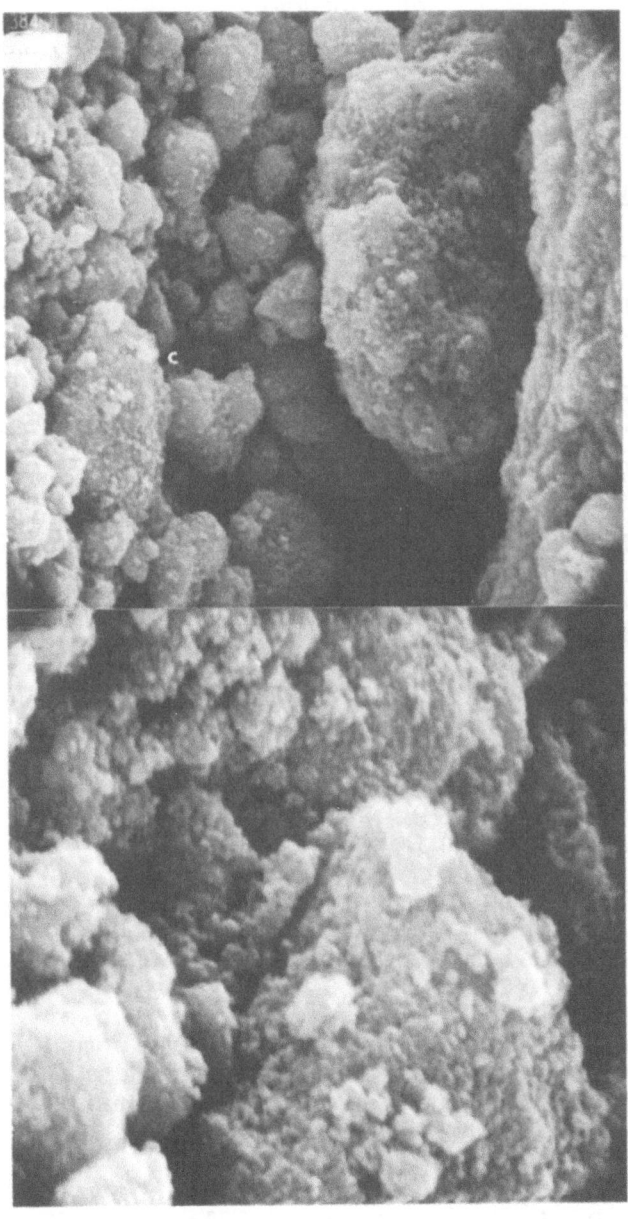

Fig. 8(c)   Top, Mag. 3000X
Fig. 8(d) Bottom, Mag. 10,000X

REFERENCES

1. Scanning Electron Microscopy/1968, O. Johari
    (ed.), IIT Research Institute, Chicago, Ill.,
    April 1968.

2. Scanning Electron Microscopy/1969, O. Johari
    (ed.), IIT Research Institute, Chicago, Ill.,
    April 1969.

3. V. Johnson, "Bibliography on the Scanning
    Electron Microscope," in ref. 2 above, p. 483.

4. O. Johari and S. Bhattacharyya, "Applications
    of Scanning Electron Microscope for Character-
    ization of Powders," J. Powder Technology, 2,
    1969, p. 335.

5. O. Johari, "Comparison of Transmission Elec-
    tron Microscope and Scanning Electron Micros-
    copy of Fracture Surfaces," J. Metals 20 (6),
    1968, p. 26.

6. W. C. Nixon, "Introduction to Scanning Elec-
    tron Microscopy," in ref. 2 above, p. 1.

7. J. C. Russ and A. Kabaya, "Use of a Nondis-
    persive X-ray Spectrometer on the SEM", in
    ref. 2 above, p. 57.

8. L. V. Sutfin and R. E. Ogilvie, Personal Com-
    munication; to be presented at 3rd Annual
    SEM Symposium, April 1970.

9. A. Boyde, "Electron Microscopic Observations
    Relating to the Nature and Development of
    Prism Decussation in Mammalian Dental Enamel,"
    Bull. Group. Int. Rech. Sci. Stomatol., 12,
    1969, p. 151.

10. S. Kimoto, M. Sato, and T. Adachi, "TV Scan-
    ning Device in a Scanning Electron Microscope
    and Its Applications," in ref. 2 above, p. 65.

THE UTILIZATION OF ELECTRON MICROSCOPY IN THE STUDY OF POWDER METALLURGICAL PHENOMENA I. NECK GROWTH MEASUREMENTS FOR SUBMICRON COPPER AND SILVER SPHERES

S. M. Kaufman, T. J. Whalen, L. R. Sefton, E. Eichen

Ford Motor Company
Scientific Research Staff
Dearborn, Michigan 48121

INTRODUCTION

The applications of electron microscopy to powder metallurgy research have, until recently, been limited in number. Previous work has been confined to the use of replicas to establish the characteristics of powder particles such as size and morphology. With continuing refinements in transmission microscopy techniques and equipment, renewed interest in the adaptation of P/M experiments to the electron microscope has arisen, specifically in the area of sintering mechanism studies. Several recent papers on the subject [1-3] have shown that standard "neck growth" experiments [4] can be performed successfully in electron microscopes equipped with heating stages. It has further been demonstrated that the growth of necks between powder particles can be followed continuously, at temperature, and linear dimensional changes in neck regions recorded with a degree of precision dependent only on the resolution of the microscope itself. This paper will describe the adaptation of the standard neck growth rate study to submicron spheres of both copper and silver with the aid of recent advances in equipment and instrumentation. An accompanying paper will describe the results of these experiments and their interpretation.

Submicron particles are well suited for sintering studies in the electron microscope, particularly from the

25

viewpoint of matching linear dimensions of sintering
particles to available magnification and resolution
capabilities of the instrument. For example, practical
resolution limits for most microscopes allow, in
principle, neck growth measurements for pairs of particles
as small as 100 Å in diameter, assuming neck radii of
the order of 10 - 30 Å. In spite of this, these particles
are at least one order of magnitude smaller than any used
previously for such measurements.

Direct transmission microscopy is also possible for
up to a particle size of about 0.8μm. Specifically,
particles less than 0.2μm are completely transparent
to an 80 KV electron beam, while particles as large as
0.8μm are transparent in the interparticle neck regions
during early stages of neck growth. Kinetic factors
provide additional inducement for using submicron
particles. Pashley et al [5] have described mass transport phenomena for particles of these dimensions as
resembling movement of liquid droplets. Rates of this
order are certain to provide benefits by minimizing
times necessary to hold a sample at temperature. This
would lessen chances of unforseen equipment-related
failures interrupting a measurement.

Thin Film Specimens

First efforts were to produce submicron particles
by the "breaking-up" of vapor deposited films into
islands. Two types of specimen configurations were
employed for this purpose, one placing the metal film
on a "wetting" substrate and the other placing the metal
film on a "non-wetting" substrate. These are shown
schematically in Figures la and b, respectively.
The carbon outer layers in Figure la were to insure
against silver diffusion through the nickel and subsequent loss by evaporation. The outer nickel layers
in Figure 1b were used to provide mechanical support
for the carbon layers when the silver film began
degenerating into islands. This proved unnecessary
and the nickel layers were later eliminated.

The composite-film specimens were made in an
ordinary vacuum evaporator equipped for multiple
evaporations without the necessity of breaking vacuum.
Each layer was deposited in sequence on a parlodian

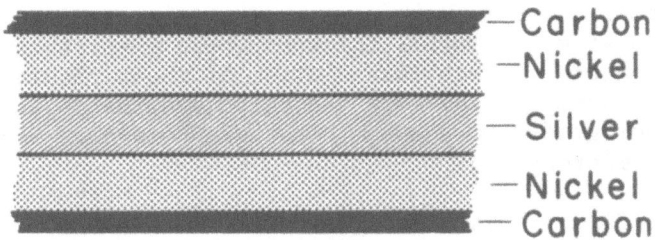

SILVER ON NICKEL SUBSTRATE

Figure 1a   Composite film specimen with Ag layer between "wetting" Ni layers.

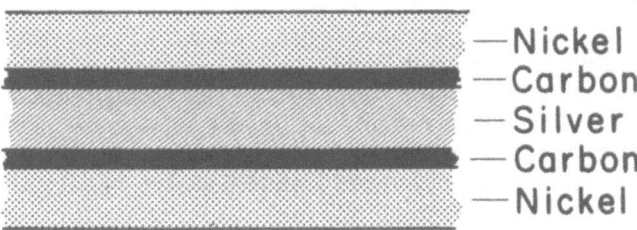

SILVER ON CARBON SUBSTRATE

Figure 1b   Composite film specimen with Ag layer between "non-wetting" C layers.

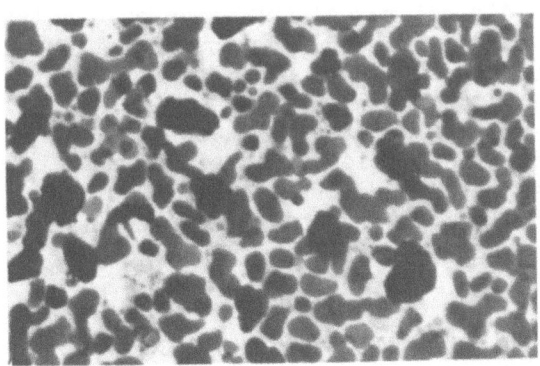

Figure 2   Silver film subsequent to heating to 200°C (10,000 X).

film supported by a glass slide. The film specimen, some
0.15 to 0.25μm in overall thickness, was then stripped
from the glass slide by dissolution of the parlodian film
in amyl acetate. Pieces of the film were then placed on
molybdenum grids for use in the heating stage of the
electron microscope.

The films, being transparent to the electron beam,
were examined by both transmission microscopy and
electron diffraction to identify microstructural
features. Heating to about 200°C caused only the silver
film to break up into islands. This occurred by a
process of simultaneous grain growth and thickening of
the larger grains followed by a fracturing at grain
boundaries forming generally polycrystalline islands.
The resultant structure was then similar to that shown
in Figure 2. No changes were observed in either the
nickel or carbon layers during the degeneration of the
silver film provided that temperature was maintained
below 300°C. For this reason experiments were conducted
at 200°C for these specimens. Migration of silver on
the nickel substrates occurred randomly and freely at
200°C. Little migration of silver on carbon substrates
was observed other than that caused by the thickening and
degeneration of the silver film itself.

Dislocations in the films of this sort tend to
line up in a direction perpendicular to the plane of
the film[5]. As such, one would not expect to observe
dislocation structures unless the film could be tilted
appreciably. Tilting of these composite film specimens
as much as 20° did not reveal any evidence of dislocation structures at any time during the formation of
islands or during neck growth events.

Numerous thermal grooves[6] were always in evidence
at triple point intersections of grain boundaries or
twin boundaries with free surfaces. It was found possible
to measure relative grain boundary energies for copper
and silver from these configurations. These results
have been published elsewhere.[7]

Neck growth measurements between island-shaped
particles were not possible with nickel substrates
because particle coalescence as normally visualized in
sintering experiments did not occur. Instead particles

degenerated into a layer-like structure and migrated by
the extension or contraction of individual layers.
Similar islands on a carbon substrate exhibited no such
tendency. The average size of particles in regions where
necks were formed on carbon substrates and coalescence
took place was about 0.01 - 0.05μm. However, reaction
or coalescence rates were so rapid that the fine details
of the process could not be captured with any of the
equipment available.

The effect of grain boundaries on neck growth in
the particles was noted repeatedly in some of the
larger particles (0.2 - 0.5μm) and re-established for
the spherical particles used subsequently. When two
particles in this size range formed a neck between them,
a grain boundary was formed in the neck also, if the
particles did not have the same relative orientation.
(If the orientation of these particles was almost the
same, coalescence occurred so rapidly that no neck
growth could be distinguished visibly.) This grain
boundary extended completely across the neck region and
caused a thermal groove to appear where it intersected
the neck surface. The local equilibrium created by
the groove configuration had the effect of pinning the
grain boundary in position.[8] For some of the smaller
particle pairs, this boundary was eliminated by re-
alignment of one of the particles. This phenomenon
was observed as an abrupt change in contrast of one of
the particles, due to reorientation, followed by an
almost instantaneous coalescence of the pair. Other
larger particles would exhibit a tendency to undergo
this type of phenomenon, but it was apparent that this
mechanism was not operative above some critical particle
size. Final coalescence between two particles was found
in all cases to depend on the elimination of this grain
boundary in the neck region. If the grain boundary
migrated from the neck coalescence could occur; otherwise, no further growth of the neck could be detected.

Spherical Particles

Production of ultra-fine spherical particles of
silver and copper was accomplished by a sublimation
technique. A recent review by Clough[9] has discussed
this subject in detail and described the production
of large quantities of submicron metal powders by

sublimation.   The process used here was performed
directly in the evacuated column of the electron microscope.   This was found to reduce unnecessary contamination of the particles from handling in air.   Direct
evaporation of silver or copper onto a cold carbon
substrate proved unsatisfactory because of the microscope maintenance problems it created.   An alternative
procedure was developed which encased the metal source
for sublimation in a carbon envelope.   These
envelopes were made by vapor deposition of a carbon
film on either side of the metal source in a manner
similar to the thin film specimens.   The metal sources
were hydrogen-reduced silver and copper powder particles
(each 99.99% pure metal) 5 to 20 microns in diameter.
A schematic representation of this configuration with
metal particles encased in a carbon envelope is shown
in Figure 3.

These particles, labelled secondary particles in
Figure 3, were produced using the microscope's electron
beam at highest intensity to create localized heating
of the large primary particle, thus causing sublimation
and condensation of the resultant vapor in cooler areas.
A wide distribution of secondary particle sizes and
shapes resulted from condensation of the vapor in the
thermal gradient around the primary particle.   The
largest particles condensed nearest the primary particle
and appeared rounded and almost spherical.   Temperatures
in these regions were probably high enough to cause a
liquid condensate.   Since neither silver nor copper
exhibit appreciable wetting on carbon[10, 11], the
approximate spherical symmetry observed for the liquid
droplets is reasonable since cooling was rapid enough
when the electron beam was turned off to quench-in
spherical shapes.   At larger distances from the primary
particle the condensate particles assumed faceted
morphologies.   This was indicative of condensation
directly to the solid state.   The faceted appearance
of these secondary particles bears strong resemblance
to the equilibrium shapes observed by Sundquist[12]
and is a direct indication of the anisotropy of surface
energy.   A typical region of secondary particles and the
original primary particle is shown in Figure 4.   The
radial distribution of particle sizes can be seen and
closer inspection will also reveal transition zones
from rounded to faceted particle shapes.

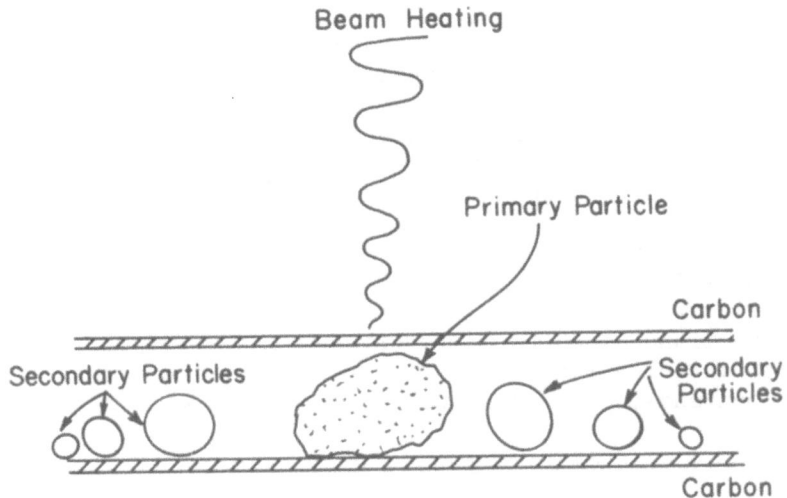

Figure 3   Preparation of spherical particles by sub-
            limation using beam heating (schematic).

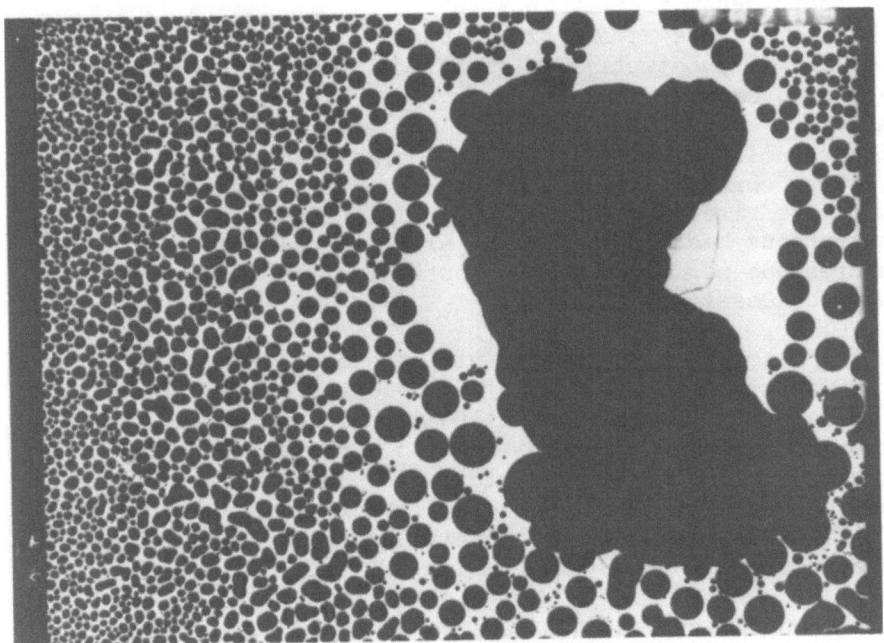

Figure 4   Representative microstructure of powder
            particle region after sublimation and
            formation of secondary spherical particles
            (10,000 X).

As mentioned before, all of the secondary particles being less than 0.8μm in diameter were transparent to the electron beam to some extent. Larger particles, in the range 0.5 - 0.8μm, were normally polycrystalline as verified by transmission micrographs near their peripheries and by the presence of thermal grooves in silhouetted micrographs of outer surfaces. Intermediate size particles (0.1 - 0.5μm) were more coarsely poly-crystalline, usually containing no more than six grains. Particles less than 0.1μm in diameter tended to be single crystals, faceted, and contained no observable crystalline imperfections. The three-dimensional nature of all of these particles was established by tilting the carbon specimen envelope ± 20° from its original position.

Sintering Experiments

The size of particles required some specialized equipment to facilitate the recording of numerical data. This equipment, a closed circuit television system, was integrated into the electron optical system of a Phillips EM 200 microscope. The television camera is shown schematically in Figure 5. This camera uses a Plumbicon tube with a fiber optics face plate and is capable of substantial image intensification. Normal electron beam currents used in these experiments were about 10^{-11} to 10^{-12} amps/cm^2, and the image was almost imperceptible to the eye on the normal flourescent viewing screen. These low beam currents combined with the size of particles under investigation made extraneous heating due to the electron beam of the microscope negligible, thus confirming the earlier work of Gessinger et al[13].

In addition to the television camera, a tape recorder and monitor were also employed. These are shown in Figure 6. Although the entire television system provided another 25X magnification, as viewed on the monitor, a decrease in resolution resulted from the video scanning lines. The practical limit of resolution for the instrument and television system was reduced to about 30 Å.

Experimental events were recorded on tape continuous-ly and were later edited for specific neck growth events. One reel of tape (3600 feet) could record continuously

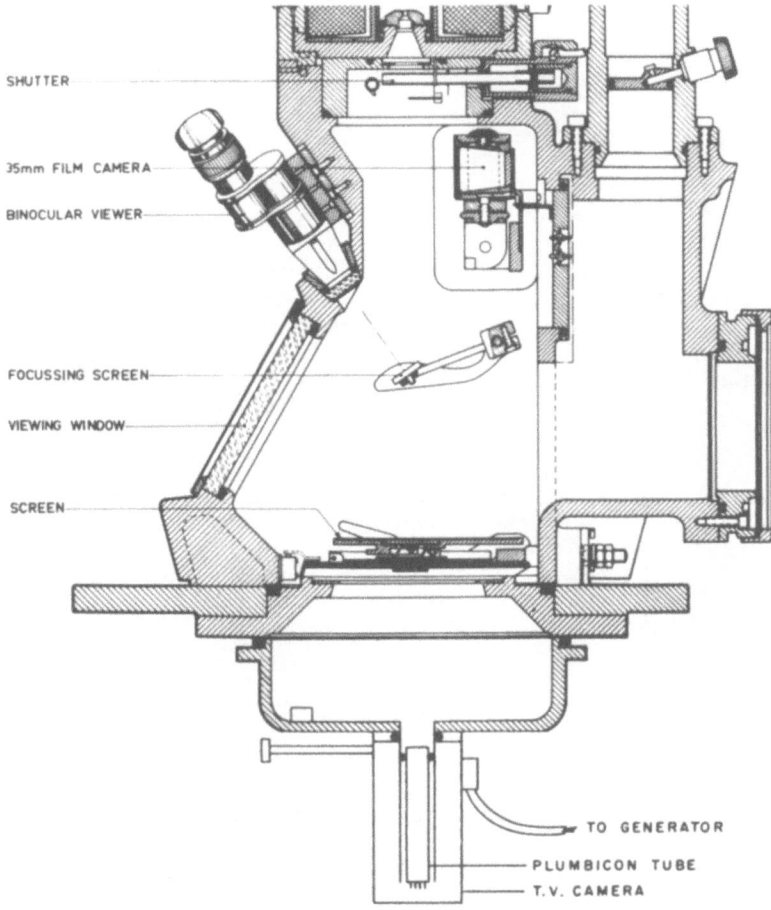

SHUTTER

35mm FILM CAMERA

BINOCULAR VIEWER

FOCUSSING SCREEN

VIEWING WINDOW

SCREEN

TO GENERATOR

PLUMBICON TUBE

T.V. CAMERA

Figure 5   Television camera adaptation to Phillips
EM 200 (schematic).

Figure 6   Video tape recorder and monitor.

for 90 minutes. Events to be retained for record and
measurement were filmed from the monitor. A kinescope
recording was made using an Arriflex 16S motion picture
camera with a "framing" bar adjustment device.
Figure 7(a) shows a frame of motion picture film with
the "framing" bar caused by lack of synchronization
between the video scanning lines and the motion picture
camera's shutter. Figure 7(b) illustrates how the
entire picture is exposed and the "framing" bar is
eliminated when the system has been synchronized.

Time scale sensitivity for the experiment was
controlled by the motion picture camera which was
capable of exposing at 30 frames/second. However,
faster shutter speeds would not have improved this
sensitivity since the video scanning lines could only
present a new image 30 times/second. In reference to
the rates of coalescence of particles observed in the
neck growth experiments, this imposed a lower size
limit of about $0.1\mu m$ diameter on the particle sizes
that could be studied. Particles smaller than this
underwent complete coalescence in times of the order of
1/30th of a second.

Actual neck growth measurements were made from the
motion picture film utilizing a Vanguard motion picture
analyzer. Early stages of neck growth were difficult
to evaluate because the limit of resolution was
approached and neck curvatures were distorted by the
video scanning lines. Figure 8 illustrates a sample
set of measurements on one pair of particles with
photographs inset to indicate the appearance of the
neck region at the times indicated. The fiduciary marks,
A and B, in the photographs were used to determine if
any center-to-center motion of the particles occurred
during neck growth. This plot is typical of both
copper and silver results in that it exhibits an
exponential dependence of neck growth with time and
little observable center-to-center shrinkage with time
until the radius in the neck region, x, is about 80%
of the particle radius, r. The irregularities in
growth rates in Figure 8 were due to grain boundaries
in the neck region causing a retardation. Inspection of
the photographs for these time intervals revealed at
least one or more thermal grooves. The slope, $n = 5.5$,
represents rates when thermal grooves and grain boundaries

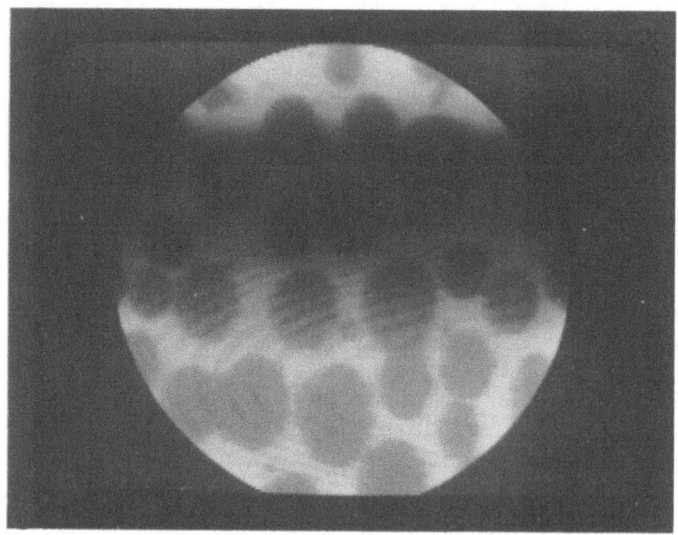

Figure 7a   Individual motion picture frame showing
"framing" bar.

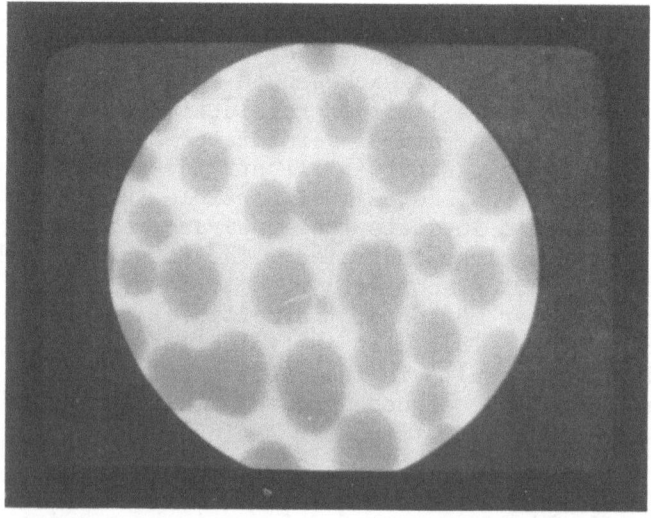

Figure 7b   Same scene taken with synchronization between
line camera and monitor image scan.

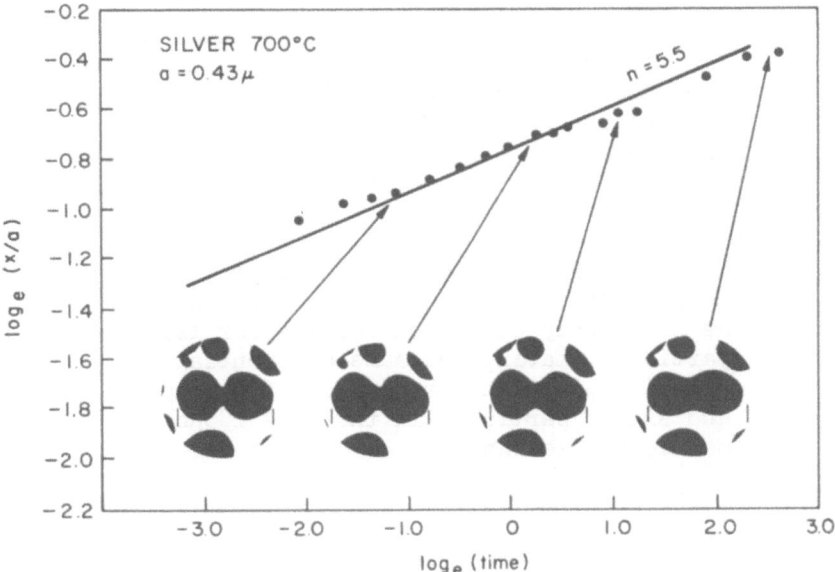

Figure 8   Typical neck growth measurements for Ag at
            700°C.

were absent from the neck region. This slope is in good agreement with the results of Kuczynski[4] for silver particles about 100μm in diameter. Equally good agreement was obtained for the copper experiments.

CONCLUSIONS

1. It has been demonstrated that standard two-particle neck growth experiments can be performed in the electron microscope with less difficulty and at least with the same precision as other existing techniques.

2. The equipment used in this investigation limited the size of particles which could be used to the range: 0.1 to 1.0μm. However, sufficient numbers of particles were produced in this size range to establish the reproducibility and accuracy of the techniques employed.

3. Rate curves generated from neck growth measurements were of the same general form as previously published results for larger particles, having the same exponential relationship with time exhibited by copper and silver particles in the 50 - 100μm particle size range.

REFERENCES

1. K. H. Olsen and G. C. Nicholson, J. Am. Cer. Soc.,
    51, 669 (1968).

2. B. E. Gere, L. Erdey, and M. Devenyi, Bany. Koh.
    Lapok (Kohaszat), 101, 319 (1968).

3. G. H. Gessinger, F. V. Lenel, and G. S. Ansell,
    Trans. Am. Soc. Met., 61, 598 (1968).

4. G. C. Kuczynski, Trans. Am. Inst. Min. Metall. Eng.,
    185, 169 (1949).

5. D. W. Pashley, M. J. Stowell, M. H. Jacobs, and
    T. J. Law, Phil. Mag., 10, 127 (1964).

6. W. W. Mullins, J. Appl, Phys., 28, 333 (1957).

7. L. R. Sefton and S. M. Kaufman, Proceedings of
    Election Microscopy Soc. of Am., 26th Annual
    Meeting, 432 (1968).

8. W. W. Mullins, Acta. Met., 6, 414 (1958).

9. P. J. Clough, "New Types of Metal Powders,"
    H. H. Hausner Ed., Met. Soc. of AIME Conferences,
    V 23, Gordon and Breach Publishers, New York,
    New York (1963), 9.

10. Yu. V. Naidich and G. A. Kolesnichenko, Poro-
    shkovaga Met., Akad. Nauk, Ukv.SSR, 6, 97
    (1966).

11. P. G. Shewmon, Trans. Am. Inst. Min. Metall. Eng.,
    227, 400 (1963).

12. B. E. Sundquist, Acta. Met., 12, 67 (1964).

13. G. H. Gessinger, F. V. Level, and G. S. Ansell,
    J. Appl. Phys., 39, 5593 (1968).

THE UTILIZATION OF ELECTRON MICROSCOPY IN THE STUDY OF POWDER METALLURGICAL PHENOMENA II. THE DEDUCTION OF NECK GROWTH MECHANISM FROM RATE DATA FOR SUBMICRON COPPER AND SILVER SPHERES

S. M. Kaufman, T. J. Whalen, L. R. Sefton

Ford Motor Company
Scientific Research Staff
Dearborn, Michigan 48121

INTRODUCTION

While the optimization of sintering rates for any material must be the ultimate objective of sintering studies on that material, little can be achieved in this direction unless some basic knowledge of the predominant mass flow mechanisms exists. A frequently employed experimental means for determining sintering mechanisms in solids has been the measurement of the rate of growth of a neck between two similar bodies during sintering. If these bodies are regular geometric forms (spheres, cylinders, etc.) the mechanisms of material transport responsible for neck growth can be deduced from absolute rate data and the temperature dependence of rate constants.

Neck Growth Rate Equations

For the purposes of this investigation, the only particle geometry which need be considered is that shown in Figure 1 for two spheres sintering together. Kuczynski[1] has shown that, for $x \leq 0.3a$ and $\rho \approx x^2/2a$, the variation of the radius of curvature of the neck, x, with time, t, can be described by

$$\frac{x^n}{a^m} = K(T)t. \qquad (1)$$

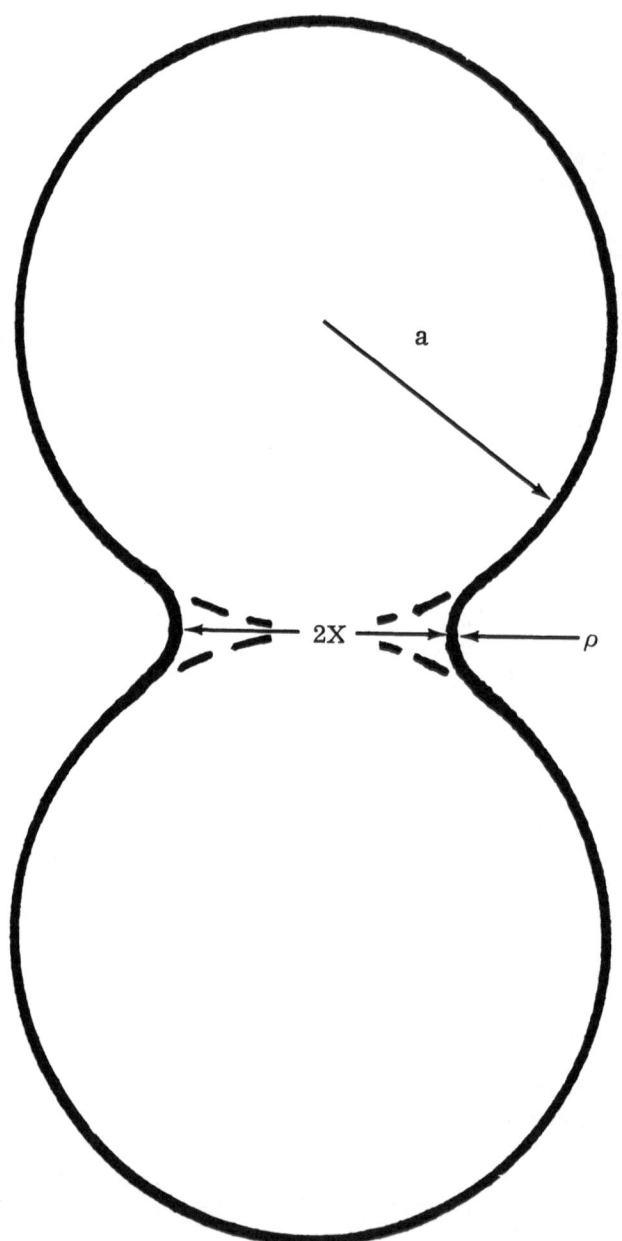

Figure 1   Geometry of the Two-Sphere Model for Sintering.

$K(T)$ is a function of temperature, T, and the mechanism of mass transport. Identification of mass flow mechanisms is accomplished by evaluation of the exponent n from a plot of log (x) or log (x/a) vs. log (time). Specifically:

$n = 2$ for viscous flow
$n = 3$ for evaporation-condensation
$n = 5$ for volume diffusion
$n = 7$ for surface diffusion

according to Kuczynski. Published results for both copper and silver neck growth experiments do not support either the viscous flow or evaporation-condensation mechanisms and, therefore, will not be considered here. Controversy exists, however, concerning the interpretation of existing copper and silver data in terms of the remaining two mechanisms. The difficulty encountered rests in distinguishing unambiguously between volume and surface diffusion mechanisms for mass flow.

The relationship given by Kuczynski governing neck growth by volume diffusion was:

$$x^5 = \frac{8\pi}{3} \left[\frac{\gamma \Omega_o a^2}{kT} \right] D_v t \qquad (2)$$

where γ is the surface tension, Ω_o is the atomic volume, D_v is the self diffusion coefficient, a is the lattice spacing, and k is Boltzmann's constant. The corresponding relationship given for neck growth by surface diffusion was

$$x^7 = 28a^2 Bt \qquad (3)$$

where $B = D_s \gamma \upsilon \Omega_o^2/kT$, D_s is the surface diffusivity, υ is the number of diffusing species per unit surface area and is approximated by $\upsilon \approx \Omega_o^2/3$. The exponent, n, in Equation (1), evaluated from experimental data usually is about 5.5 for both copper and silver. Distinguishing between exponents 5 and 7 in Equation (2) and (3) is possible with data that are reasonably precise. Recently, however, Nichols and Mullins[3] using a more exact treatment for the geometry of the neck region in the two-sphere model have shown that the exponent, n,

in Equation (3) actually varies between 5.5 and 6.5 as
x/a increases and is better approximated for x/a < 0.4
for surface diffusion by

$$x^6 \approx 25a^2 \, Bt \qquad (4)$$

The difficulty in distinguishing between volume and
surface diffusion becomes much more acute when Equation
(4) is considered in place of the Kuczynski 7th power
relationship for neck growth. Some investigators have
attempted to resolve this difficulty by calculating D_s
from Equation (4) or D_v from Equation (2), depending on
which of these appeared to approximate the experimental
results more closely. Unfortunately such a calculation
does not usually resolve the question of which of these
mechanisms is operative, as illustrated in the following
examples:

Based on Kuczynski's data for silver at 800°C[1],
Kuczynski calculated D_v = 1 x 10^{-9} cm^2/sec. while
Johnson and Clark[4], using a correction term for grain
boundary contributions to diffusion with this same data,
obtained D_v = 4.8 x 10^{-10} cm^2/sec. These are to be
compared to the measured value of D_v = 4.2 x 10^{-10} cm^2/sec.
of Hoffman and Turnbull[5]. The good agreement was noted
and both of these studies concluded that volume diffusion
must be the mechanism for neck growth in silver. However,
using the same basic data of Kuczynski, Nichols[2] employed
Equation (4) to calculate the temperature dependence of
D_s and found:

$$D_s = 2.5 \text{ x } 10^5 \exp\left[\frac{-49,700}{RT}\right]$$

as compared to the measurements of Rhead[6] which have
shown

$$D_s = 10^6 \exp\left[\frac{-57,000}{RT}\right].$$

It should be obvious at this point that the origin of
most of the controversy in this area lies in the
limitations of the neck growth equations themselves.
Considering the degree of precision which can reasonably
be expected in experiments of this type, the Kuczynski
equations are not applicable to sufficiently large

time intervals to allow a distinction to be made between
the 5th and 6th power relationships in x.

Herring's "Scaling" Laws[7]

The "scaling" laws were also designed to assess
the relative contributions of the four primary mass
transport mechanisms to the sintering process, in this
case by a comparison of the changes in time scale with
changes in particle size for particle configurations
possessing geometrically similar shapes. As stated by
Herring, they are of the form:

$$\Delta t_i = \lambda^n \Delta t_1 \qquad (5)$$

and describe the time necessary, Δt_i, for some particle
configuration (i) to reach a stage of sintering identical
to that reached by another particle configuration (1) in
a time Δt_1; the two configurations differing only in scale
of linear dimensions. This size difference is described
by λ, the ratio of linear dimensions of particle
configuration (i) to particle configuration (1). The
exponent n has a value 1 for viscous flow, 2 for evaporation, 3 for volume diffusion, and 4 for surface
diffusion. The primary assumption in this formulation
is that the mechanism of sintering for any given material
at a given temperature is not size sensitive. There is
some evidence to the contrary[8, 9] in the literature
indicating that more complex mechanisms may become
operative at very large particle sizes (> 300 microns).
However, recent work[10] has confirmed that in the range
of particle sizes of 10 - 300 microns, the scaling laws
are valid for copper and most likely for silver also.

Deduction of mechanisms of mass transport from the
"scaling" laws is subject to controversy also because
published experimental results do not cover a large
enough range of particle sizes to allow differentiation
between a 3rd or 4th power relationship of λ, the scale
factor. The importance of the study of neck growth rates
in particles less than one micron in diameter is thus
defined clearly as the means by which to overcome
insufficiencies in the two primary theoretical treatments of sintering rates in the treatment of experimental
results.

Although much of the subject of powder metallurgy
deals with situations where particle sizes are less than
one micron in diameter, very little work, other than
phenomelogical studies on powder compacts, has been
published. This has undoubtedly been due to the lack
of experimental tools capable of resolving small neck
regions. Recently the electron microscope has been
adapted to neck growth studies for copper[10], gold[11],
and silver[12], although not in this size range. The
work by Gessinger, Lenel, and Ansell[12] includes
observations for particles as small as 5 microns.
Contamination of their samples in the microscope caused what
neck growth that was observed to cease completely and
made meaningful rate measurements impossible. The work
to be described here was a natural extension of these
previous studies, and use was made of somewhat more
refined experimental equipment and techniques. These
refinements permitted continuous observation, at
temperature, of neck growths for both copper and silver
particles less than one micron in diameter.

RESULTS

Neck Growth Rates as a Function of Temperature

Most of the experiments were performed at 700°C.
This was found to be an optimum temperature for both
copper and silver from the standpoint of the measurement of time and linear dimensions. Lower temperatures
permitted finer resolution of time scales for particles
less than 0.1 micron diameter, but resolution of neck
dimensions at early times was beyond the capabilities
of the microscope-television system. Conversely high
temperatures increased the neck growth rates of particles
with resolvable neck regions until the resolution of the
time scale was no longer adequate. Even though a range
of particles sizes (0.4 - 1.0µm) existed that would
tolerate investigation of the temperature dependence of
neck growth rate, there was enough uncertainty in
temperature measurement to make any conclusions that
might be drawn subject to question. Some work was done
at 900°C with the few large secondary particles available but this was limited because of the inability to
create a sufficient number of interparticle necks. This
was believed to be due to a lower concentration of
particles per unit area, and thus a lowering of the

probability of finding two particles close enough to form
a neck. Too few examples were observed to draw any
conclusions other than the obvious increase in neck growth
kinetics as temperature was increased.

Morphological Changes

In Part I of this paper, it was illustrated that
particles less than about 0.1 micron are predominantly
single crystals as condensed from the vapor. Additional
thermal treatment appeared to have little or no effect
on the faceted morphology of these particles. Only when
two or more of these particles coalesced did any rounding
of the faceted outer surface take place. This, in most
cases, was only temporary and the resultant larger
particle, also a single crystal, assumed a similar
equilibrium shape. The larger polycrystalline particles
exhibited tendencies to facet with little or no motion
of grain boundaries. Thermal grooves, when observed,
remained static during the entire faceting process within
the limits of dimensional resolution. Faceting occurred
much more rapidly at the higher temperatures. No
significant difference was observed between the times
necessary for silver or copper particles not in contact
to form facets. Again, for the larger particles, facets
remained stable until particles came in contact where-
upon rounding occurred rapidly. The material displaced
in this process presumably migrated into the neck region
and acted initially, at least, as the primary source of
mass for neck growth. During this rounding, grain
boundaries were observed to become more mobile and could
migrate in and out of neck areas causing some discontinuities in neck growth rates.

Creation of Interparticle Necks

The process of creation of interparticle necks was
observed many times for both silver and copper. It
was perhaps the most difficult of all observations to
explain. Prior to the appearance of a definite neck, a
grey area somewhere in the region of closest approach of
the particles becomes apparent. The appearance of the
grey area was found to be necessary but not sufficient
to establish contact since in many instances the
particles would not join. This phenomenon was observed
only at temperature and was established experimentally

not to be an electron optical effect, at room temperature
at least, by tilting the specimen with the goniometer
stage. There was no evidence of crystallinity observable
in these interparticle grey areas so that it could not
be established whether or not material transport was actually occurring. The only evidence as to the mechanism
of establishing interparticle contacts was indirect and
may be inconclusive. Experiments performed for tin were
unsuccessful in that no contacts could be established
even 200 to 300°C above the melting point. If the
surfaces were equivalent in all cases examined, one
could conclude that vapor transport was important in
establishing contact since tin has a very low vapor
pressure at these temperatures relative to both copper
and silver. It is also highly possible due to the
irregular nature of the carbon substrates, that contact
may be established mechanically, by gravity for example,
and that the grey area observations may be a characteristic of the microscope-television system.

Neck Growth in Silver Particles Smaller than 0.1μm

Particles substantially less than 0.1μm diameter
coalesced completely in 1/15th of a second or less. A
typical example of this is shown in Figure 2. The three
pictures were taken from three consecutive frames of
motion picture film and were taken 1/30th of a second
apart. The frame immediately before 2(a) showed a clear
separation between the two particles. The frame in 2(a)
shows the grey area referred to previously in the
establishing of interparticle contact. Figure 2(b) was
taken after much of the coalescence had occurred and
illustrates the difficulty in stopping action because of
the television and motion picture system limitations on
time increments. Figure 2(c) shows the coalescence
complete with only the "ghost" of an image of the original
particle in the upper part of the photo. It is evident
from this sequence that deduction of mass transport
mechanisms for particles such as these was beyond the
capabilities of the system. Some evidence can be cited
to support the contention that events of the type shown
in Figure 2 are neck growth governed primarily by
surface diffusion, at least in the early stages. Pashley
et al[13] have shown, using Kuczynski's neck growth
relations[1] that the times expected for 0.1μm diameter
particles of gold in contact to achieve a neck diameter

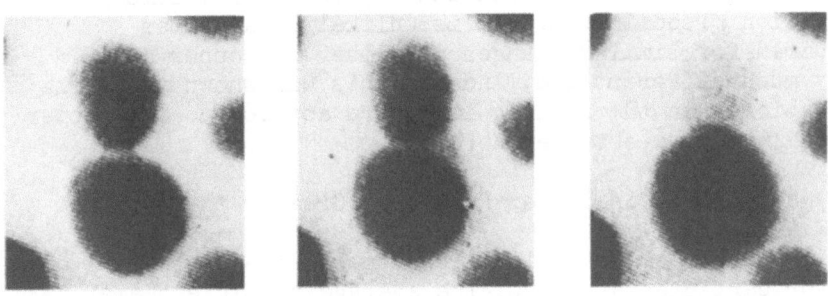

SILVER 700°C

Figure 2   Coalescence of Ag particle pair at 700°C
            (a = 0.01μm).

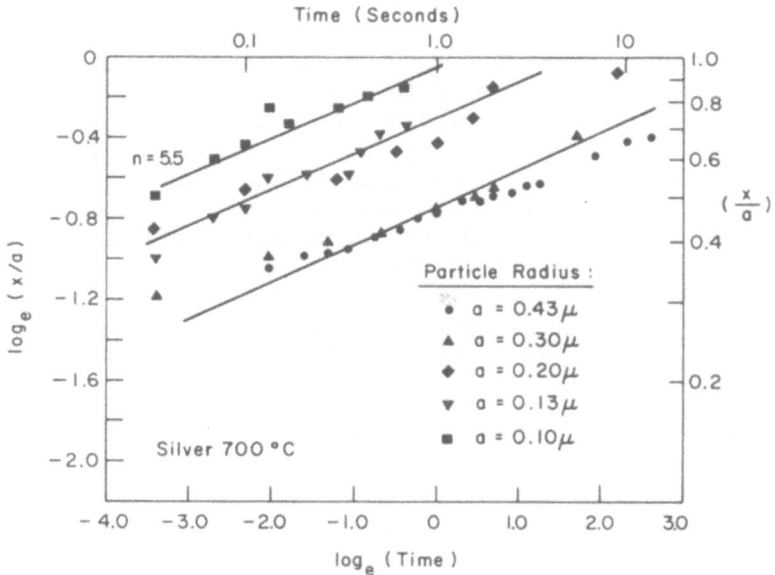

Figure 3   Neck Growth Data for Ag at 700°C.

of 0.1μm would be 10^{-3} seconds if surface diffusion was
responsible for neck growth and 2 seconds if volume
diffision predominated. It is unlikely that times
required for similar changes in silver or copper are
very much different and, indeed, this was shown to be the
situation when the scaling laws were applied to the copper
and silver data obtained here.

Neck Growth in Silver Particles Greater than 0.1μm

Figure 3 shows typical results obtained for pairs of
silver particles at 700°C, the average particle diameters
varying from 0.2 to 0.86 microns. Superimposed on this
plot is a series of lines drawn with an inverse slope of
5.5, an average of reported values from other investigations. Two significant facts may be drawn from this plot.
First, it appears from the linearity of the experimental
points for any given particle pair that the same mechanism
is operative throughout the entire interval of (x/a) shown.
Secondly, the slope of these lines appears to be approximately in the same range normally observed for x/a ≤
0.3. The scatter in the experimental results does not
allow a legitimate least squares analysis, nor does it
allow a claim that the average of "best fit" slope of
these lines is 5.5 as compared to 5 or 6. Normal
scatter due to measurement errors does occur because of
the minimum time increment being fixed at 0.033 sec. and
because some of the small neck radii of the order of
100 Å are beginning to approach the limits of resolution
of the microscope. These errors are of the type that
can be normalized by a "least squares" analysis.
However, a third source of scatter in the data was
found to be due to the irregular nature of growth of these
interparticle necks when a grain boundary was present in
the neck region.

Copper Results

The results for neck growth measurements for pairs
of copper particles were essentially similar to those
for silver at 700°C. The only exception to this was
that the range of particle sizes obtainable for copper
was not nearly as large. Difficulties were encountered
producing enough particles near one micron in diameter to
study larger particle pairs. As a result the particle
range for investigation was reduced to 0.1 < a < 0.5

microns. The data for copper at 700°C are shown in Figure
4 along with the lines whose inverse slope is 5.5. The
erratic nature of the lower set of data points for a =
0.25μm was due to presence of grain boundaries in the
neck region for this pair of particles. Other than this
one area which exhibits an unusually large degree of
scatter, it is evident from the linearity of the data
points that observations similar to the ones made for
silver spheres hold also for the case of copper spheres.

DISCUSSION

Effect of Grain Boundaries

     The effect of grain boundaries on neck growth which
was observed in these experiments was contrary to what
has been cited previously in the literature for either
copper or silver. Using data for particles much larger
(~ 100μm) Ichnose and Kuczynski[14] and more recently,
Johnson and Clark[4] have treated the subject of the
influence of grain boundaries on neck growth rates. The
conclusion arrived at by both of these treatments with
respect to rates is that, if volume diffusion were the
operative mechanism for neck growth, the presence of a
grain boundary would be expected to enhance the rate of
neck growth. The grain boundary is viewed in this model
as a sink for vacancies migrating from the free neck
surface and thus provides a rapid diffusion path. However,
if the surface diffusion were the operative mechanism, the
situation becomes reversed. Wilson and Shewmon[10] have
shown that the grain boundary diffusion rate should be
negligible compared to that of surface diffusion where
temperature approaches to within 70% of the melting point.
Further, since the driving force for surface diffusion
arises from local variations in curvature of the free
neck surface, the formation of thermal grooves at the
intersections of grain boundaries with the neck surface
can effectively halt the flux of surface atoms in these
regions by removing the driving force. On the basis of
these arguments, the observed effect of grain boundaries
in this study suggests that a surface diffusion mechanism
is operative. Gessinger et al[13] may also have observed
this same effect occurring in larger particles of silver,
some 5 - 30μm in diameter. They reported cessation of
neck growth in several instances after an initial rapid
thickening of the neck. Although this was attributed to

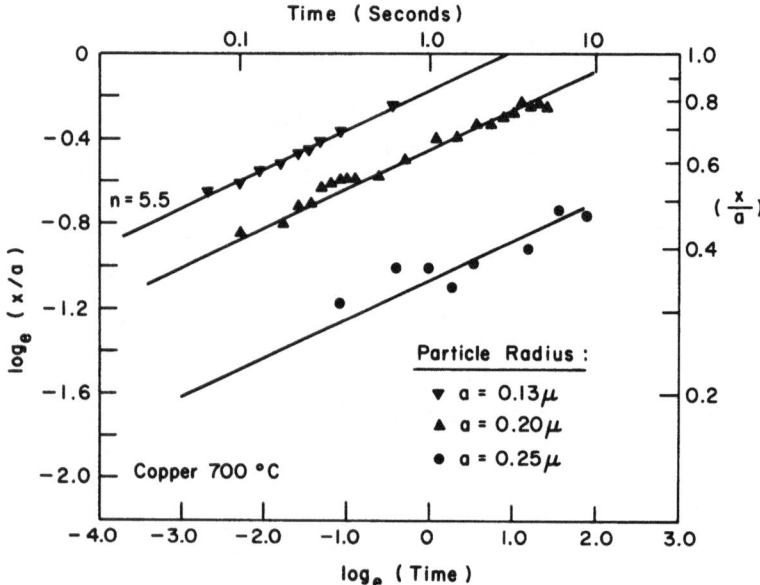

Figure 4   Neck Growth Data for Cu at 700°C.

build-up of contaminants on the particle surfaces,
inspection of the published electron-micrograph does
indicate the presence of thermal grooves in interparticle necks.

Surface Diffusion as a Model for Sintering Mechanisms

It has become increasingly evident as more and more
work in this area appears in the literature that controversy
exists over whether volume diffusion or surface diffusion
controls neck growth. The models proposed for volume
diffusion require that sintering which is volume diffusion
controlled must result in a volume decrease or shrinkage
of the particle array. In the geometrical configuration
employed here, this would correspond to a decrease in
center-to-center distance for the particles.   This was
not observed until very late in the coalescence process
and, as such, volume diffusion in the early stages of
neck growth does not appear to be the operative mechanism.

Considerable indirect evidence supporting a surface
diffusion controlled sintering process in these materials
has already been cited. Direct deduction of sintering
mechanism from the data as presented up to this point was
not possible. The measurements made for $x/a \leq 0.3$ were
obtained for times varying from 1/30 to 5/30 seconds.
Errors in time measurement of one frame of motion picture
film, 1/30 seconds, introduced too much uncertainty to
make the results obtained reliable enough to utilize
the Kuczynski equations. Similarly, the data were not
sufficiently precise to allow application of Herring's
"scaling" laws to differentiate between the possible
mechanisms. This is shown in Figure 5. Here Equation
(5) has been used to reduce all the results for silver to
the time scale of 0.8 micron diameter particles. The
individual points have been eliminated on all four of
these plots and the cross-hatched areas substituted to
show the spread in the calculated results. Ideally, if
the "scaling" laws indicated the operative mechanism, one
of these plots would exhibit a significantly smaller
amount of scatter than the other three. Evaporation-
condensation and viscous flow processes can be eliminated
from consideration relative to the diffusion processes
on the basis of these plots. As in the case of previous
investigations, the unambiguous distinction between
surface and volume diffusion is not present.

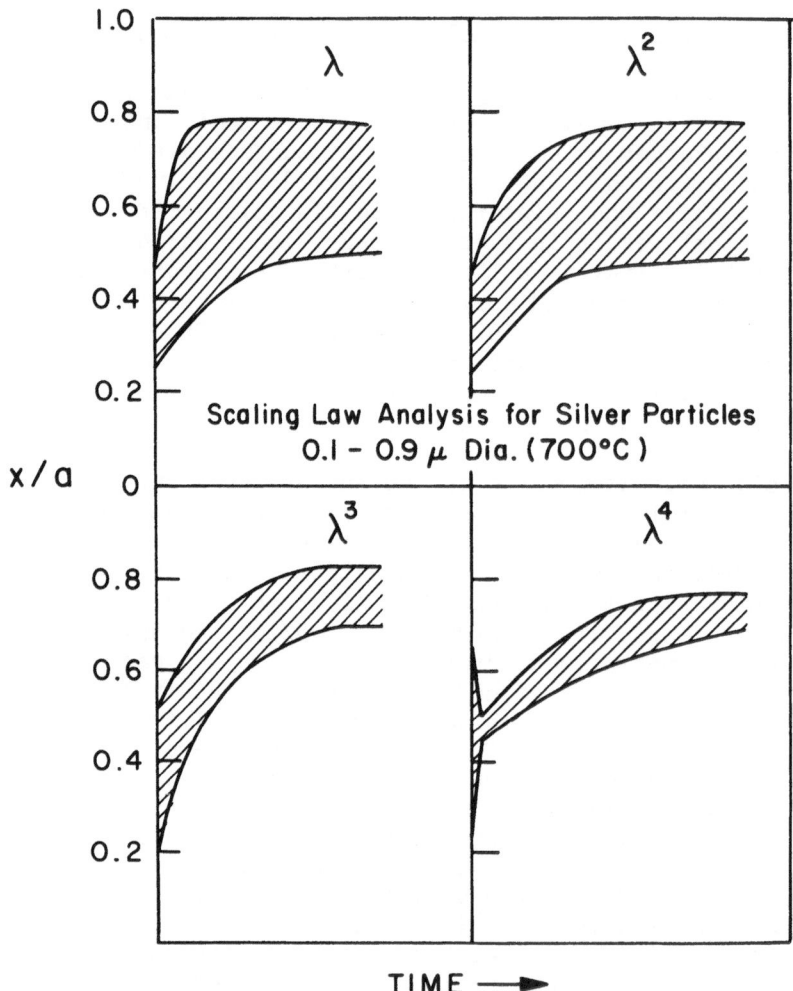

Figure 5   Ag Neck Growth Data "Scaled" to a = 0.4μm.

To confirm directly that surface diffusion is the mechanism of mass transport which is operative in these systems of small particles, the "scaling" laws were applied to data on all particle sizes available for the two-sphere geometry. For this purpose Equation (5) was reduced to the form

$$\frac{\Delta t_i}{\Delta t_1} = \lambda^n$$

and a general plot constructed, as in Figure 6, for n = 3 and n = 4 for a one decade interval of time. The cross-hatched bands indicated the reduction in magnitude of this decade in time as particle size is decreased two orders of magnitude. If the "scaling" laws are assumed valid for particles in the size range a = 0.1μm to a = 100μm, then whatever mass flow mechanism is operative at λ = 1 will also be operative at λ = 0.01. Assume configuration (1) is chosen such that a_1 = 50μm; the distinct separation of the bands at λ = 0.01 indicates that whichever of the two mechanisms is operative can be unambiguously identified by comparison of data for particles about 100μm diameter to particles about 1μm in diameter. As an example, by extrapolation of Wilson and Shewmon's data for copper[10], at 700°C the decade of time including 100 - 10,000 hours corresponds to x/a varying from about 0.3 to 0.4 for the 100μm particles.* For comparrison to one micron particles λ = 0.01. If volume diffusion controls the neck growth in this size range then

$$\frac{\Delta t_i}{\Delta t_1} = 10^{-6} - 10^{-5} \text{ for x/a varying from 0.3 to 0.4}$$

The lower limit on Δt_1 for x/a = 0.3 is 1000 hours and thus Δt_i becomes

$$\Delta t_i = 3.6 \text{ seconds}$$

for this lower limit and the decade of time for this process by a volume diffusion mechanism becomes 3.6 to 36 seconds. Similarly, the calculation for surface

---

* Kuczynski's[1] data show similar geometric changes for silver particles and do not differ appreciably in time interval.

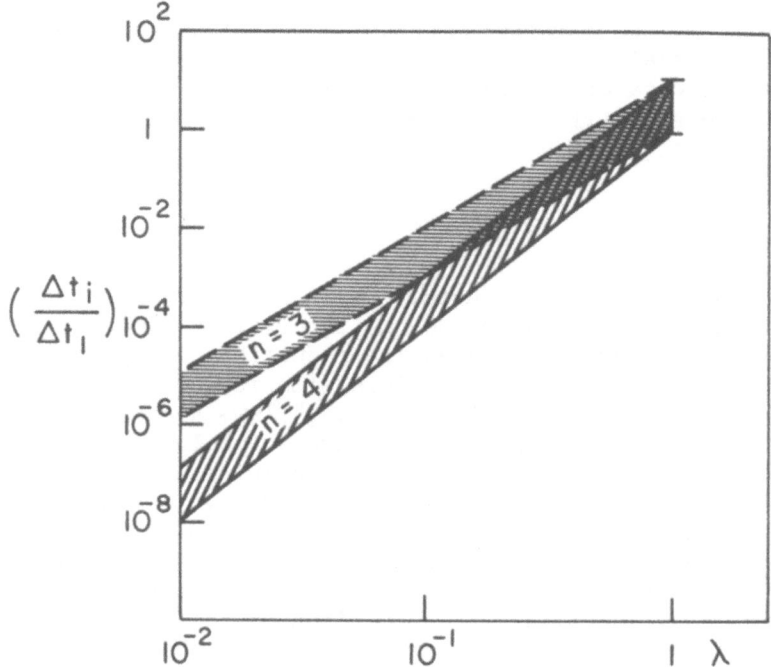

Figure 6   Relative Change of Time Scale with Change in
            Particle Size for Volume (n = 3) and Surface
            (n = 4) Diffusion. λ = scale factor.

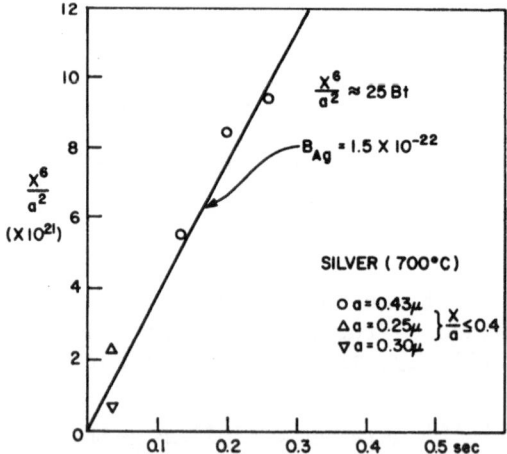

Figure 7   Comparison of Ag Data for x/a ≤ 0.4 with
            Empirical Values for the Parameter B.

diffusion controlled neck growth reduces the decade of time
to the range 0.036 to 0.36 seconds. This is the order of
change that was experimentally observed here, occurring
in one to three frames of motion picture film, or 0.033 –
0.099 seconds, for particles about 0.8µm in diameter.
It is clear, therefore, that volume diffusion cannot be
responsible for the rapid changes in neck diameters
observed here, the predicted rates being two orders of
magnitude too slow. The reasoning leading to deduction
of a surface diffusion mechanism for neck growth in silver
is analogous since the observed kinetics at 700°C were
almost identical. Gessinger et al[12], in their only
reported rate measurement showed that 10 micron silver
particles at 615°C reach $x/a = 0.3$ in times of about
30 to 40 minutes. Figure 6 predicts one hour for volume
diffusion and about 6 minutes for surface diffusion at
700°C. Reducing temperature to 600°C causes a 2.2
multiplication of the time scale[1], and thus Δt_i becomes
2.2 hours for volume diffusion and 13 minutes for surface
diffusion controlled neck growth. Once again, the volume
diffusion model predicts rates considerably slower than
those observed experimentally.

     The Nichols and Mullins equation [3] describing
neck growth between spheres by surface diffusion can be
applied here. Since few data points were obtained in the
region $x/a < 0.4$, as required by the Nichols and Mullins
treatment, some justification is needed to extrapolate
the lines in Figures 3 and 4 to log (time) = 0. Figure
7 provides the evidence that this linear extrapolation
is indeed proper.   The line shown was calculated from
the Nichols and Mullins equation using known data for the
surface energy[15] and self-diffusion constant for silver[6]
The points shown were measurements made from the submicron
particles. The agreement does suggest that inverse slopes
of 5.5 can be assumed for the data obtained for $x/a < 0.4$.
The calculation of values of D_s, from Equation (4) was
performed for both copper and silver at 700°C. These
values are shown in Table I along with results of other
sintering investigations for larger particle sizes and
the actual experimental measurements. Once again
agreement with actual measurements of D_s is at least as
good, in one case better, than obtained by previous
investigations of this type.

TABLE I

CALCULATION OF D_s FOR SILVER AND COPPER FROM THE
KUCZYNSKI NECK GROWTH EQUATIONS (700°C)

Particle Radius: a (microns)	n	D_s (cm^2/sec) Calculated	Measured	Material
0.1	5.5	8.0×10^{-8}		Silver
0.2	5.5	8.5×10^{-7}		Silver
0.3	5.5	1.1×10^{-7}	3×10^{-7} (6)	Silver
0.43	5.5	4.0×10^{-7}		Silver
60.0	5.65	5.0×10^{-6} (1)		Silver
0.13	5.5	1.1×10^{-6}		Copper
0.20	5.5	1.5×10^{-6}	8×10^{-7} (16)	Copper
3 – 20	6.5 (10)	1×10^{-6}		Copper
100	5.3	–		Copper

It may appear from what has been presented that
measurements made for small-particle neck-growth rates
are not sufficient, in themselves, to establish sintering
mechanisms. This is not so, and this can be shown from
Figure 6. Suppose that data had not been available from
larger particle sintering experiments. The band of values
shown for n = 3 could alternatively have drawn so as to
originate from the same interval as n = 4 at λ = 0.01,
that is, moved vertically two orders of magnitude so that
the one micron particle experiments were the reference
configuration. This analysis would thus predict that, if
volume diffusion were the mechanism of neck growth, the
interval of time required for the neck to grow from x/a =
0.3 to x/a = 0.4 would be 10 to 100 hours. The implied
rapid rate of neck growth according to this prediction
could then have been easily shown to be incorrect just
on the basis of sintering experiments for other metals
at comparable temperatures below their melting points.
Normal times observed for this change would be at
least one order of magnitude greater.

Later Stages of Particle Sintering

The mechanism or mechanisms by which the ultimate coalescence of particle pairs occurs is not clear from the results obtained here. In addition, the mechanism by which a grain boundary, supposedly trapped by a thermal groove, migrates rather rapidly and abruptly out of the neck region is not clear either. The coalescence, because of a center-to-center distance decrease for the particle pair, must occur by either volume diffusion or plastic flow or some combination of these mechanisms. Without appropriate theoretical models, however, it was not possible to distinguish between these mechanisms.

CONCLUSIONS

1. The electron microscope provides a useful and rapid method for determination of sintering and mechanisms for particle sizes in the region of most interest to powder metallurgists.

2. For both copper and silver at 700°C, it has been established that neck growth occurs between spherical submicron particles by a surface diffusion mechanism.

3. The effect of grain boundaries has been shown to be opposite to that observed in cases where sintering is volume diffusion controlled. Grain boundaries trapped in interparticle necks caused retardation or cessation of neck growth.

4. Calculated values of the surface diffusivities of copper and silver at 700°C were in good agreement with published experimentally determined values.

REFERENCES

1. G. C. Kuczynski, Trans. Am. Inst. Min. Metall. Eng., 185, 169 (1949).

2. F. A. Nichols, Acta. Met., 16, 103 (1968).

3. F. A. Nichols and W. W. Mullins, J. Appl. Phys.,
    36, 1826 (1965).

4. D. L. Johnson and T. M. Clark, Acta. Met., 12,
    1173 (1964).

5. R. E. Hoffman and D. Turnbull, J. Appl. Phys.,
    22, 634 (1951).

6. G. E. Rhead, unpublished research as quoted by
    N. A. Gjostein in "Metal Surfaces: Structure,
    Energetics and Kinetics", p. 150, Am. Soc. Met.,
    Metals Park, Ohio (1963).

7. C. Herring, J. Appl. Phys., 21, 301 (1950).

8. J. G. R. Rockland, Acta. Met., 15, 277 (1967).

9. R. T. DeHoff, D. H. Baldwin, and F. N. Rhines,
    Plans. F. Pulvermetall., 10, 24 (1962).

10. T. L. Wilson and P. G. Shewmon, Trans. Am. Inst. Min.
    Metall. Eng., 236, 48 (1966).

11. K. H. Olsen and G. C. Nicholson, J. Am. Cer. Soc.,
    51, 669 (1968).

12. G. H. Gessinger, F. V. Lenel, and G. S. Ansell,
    Trans. Am. Soc. Met., 61, 598 (1968).

13. D. W. Pashley, M. J. Stowell, M. H. Jacobs, and T.
    Shaw, Phil. Mag., 10, 127 (1964).

14. H. Ichinose and G. C. Kuczynski, Acta. Met., 10,
    209 (1962).

15. F. H. Buttner, E. R. Funk, and H. Udin, Phys. Chem.
    Solids, 56, 657 (1952).

16. H. P. Bonzel and N. A. Gjostein, Phys. Stat. Sol.,
    25, 209 (1968).

A BUBBLE RAFT MODEL TO STUDY SINTERING BY PLASTIC FLOW

F. V. Lenel, G. S. Ansell, and R. C. Morris

Materials Division
Rensselaer Polytechnic Institute
Troy, New York 12181

I. Introduction

The geometrical changes observed during the sintering of solid particles can be accounted for by a variety of atomic mechanisms, including various diffusional fluxes implicit in the geometry of the model systems, and shear processes involving pure slip of dislocations and climb or cross-slip controlled creep.

A body of experimental evidence exists which indicates that a portion of the neck growth and shrinkage observed in the sintering of soft metals and ionic solids such as Cu, Ag, Zn, CaF and ThO_2 must be ascribed to slip mechanisms (1-6). Of particular interest is the work of Nunes et al on the anisotropy of neck growth between variously oriented single crystals of zinc as a function of the anisotropy of diffusion and slip in the zinc lattice which essentially prove_ that a substantial portion of the sintering is due to slip.

The results of Morgan et al (1,2) which show a temperature dependent end point in the shrinkage of sintering CaF_2 and ThO_2 compacts, and the marker experiments of Hingorany et al (4) on the sintering of CaF_2 spheres and plates are also impressive.

A detailed theory of sintering by dislocation motion would entail a knowledge of: (1) the time dependence of the geometry and the state of stress in the sintering neck, (2) the interior and surface sources of the dislocations

61

involved and their slip systems and (3) work hardening and relaxation mechanisms. Generally, the problem is very difficult owing to the complex nature of the stress tensor at various points in the deforming regions, the multiplicity of available glide systems and the intricate sequences of dislocation generation, glide and interaction which would be required to explain the observed geometrical changes as a response to the stresses resulting from the annihilation of free surfaces. Further complications arise from the effects of simultaneous diffusion fluxes on the various stages of the slip processes. The net result is that even a qualitative picture of the process is not easily agreed upon and that attempts at quantitative analysis of the state of stress and the resulting rate of transport are of a most approximate nature and frought with questionable assumptions. The diffusional mechanisms, on the other hand, being geometrically simpler, are more easily visualized and have been mathematically modeled in some detail, which accounts in part for their wider acceptance and the prevalent mistaken neglect of slip and creep contributions to sintering.

In order to establish a more detailed qualitative picture of the shear processes involved in sintering, and thereby possibly facilitate a less dogmatic approach to the topic, the phenomenon was simulated using circular areas of 2 dimensional bubble rafts as analogeus to sintering metal particles. The analogy between the elastic and plastic properties of close packed monolayers of 1-2 mm. diameter bubbles on the surface of a liquid, initially pointed out by Bragg et al (7-9) is quite interesting and well known.

Bragg and Nye (7) showed that rafts of uniformly sized bubbles, upon the application of stress, deformed elastically and then plastically by dislocation generation and slip in a manner quite similar to a metal lattice. Calculations modified by Bragg and Lomer (8) showed the close correspondence between the attractive and repulsive forces operating in a lattice of bubbles and those calculated for copper ions in a copper lattice. Experimental measurements of the elastic constants of bubble rafts and their elastic limit as functions of bubble size demonstrated the validity of the calculations (9). It was shown that, for the surface energy and capillary constant of the soap solutions used in the experiments, a bubble diameter of 1.2 mm. yielded the closest fit to the potential curves for copper ions. The measured elastic limit of perfect rafts of

bubbles of this diameter corresponds to a shear strain of
3.25°. The maximum shear strain of dislocation free copper
whiskers has since been shown to be about 2° (10), which
illustrates the basic soundness of the analogy for non-
dynamic situations. The analogy fails, however, in the
case of the dynamics of a gliding dislocation, since the
bubbles have essentially no mass and are situated in
viscous medium. Thus, a dislocation moving in a bubble
raft has little kinetic energy and energy is dissipated by
viscous flow of the liquid, while in a metal lattice the
moving dislocation has appreciable kinetic energy and
energy is dissipated by phonon radiation. A final point,
bearing on the use of the bubble raft model to study
sintering, is that since the bubbles have no mass there is
no way to simulate thermal vibrations and the associated
thermally activated processes of diffusion and creep controlled by thermally activated cross slip and climb.

The bubble raft model should therefore provide a useful
picture of the geometrical aspects of sintering by pure
slip. The contributions to the geometry change from diffusion or thermally activated creep and the kinetics of the
slip processes are not simulated.

II. Experimental

The soap used in this experiment was the same composition used by Bragg and coworkers and is prepared as
follows:

> 60.8 cc. of reagent grade oleic acid and 200 cc.
> of distilled water are shaken together thoroughly,
> and the resulting suspension is mixed with 292 cc.
> of 10 weight % triethanolamine. Water is added
> to make 800 cc. and 656 cc. of reagent grade
> glycerine is blended in. The mixture is allowed
> to stand for 1-2 hours and the clear solution is
> drawn from the bottom and dilluted with 3 parts
> by volume of water (7).

The apparatus is shown in Fig. 1. The soap solution
passes from the reservoir through a valve and into the
glass manifold around the bottom edge of the tray. The
level of the solution is raised to 1 or 2 mm. below the
tops of the forming rings. Bubbles are blown into the
rings through a 0.2 mm. glass capillary tube connected to a

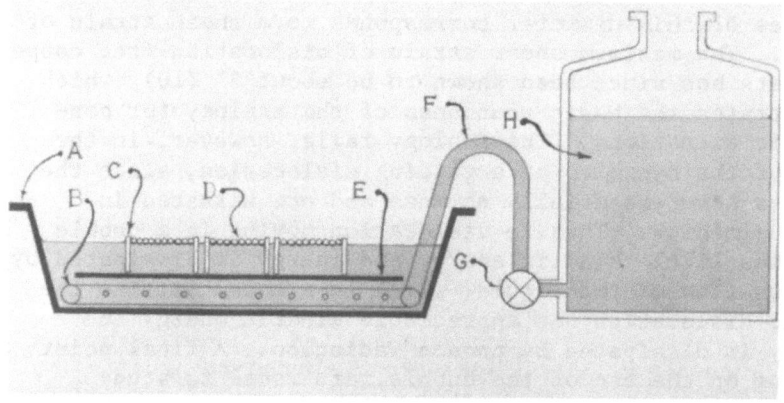

Fig. 1. Apparatus for the study of bubble raft
sintering: A. Tray, B. Manifold, C. Ring Forms,
D. Bubbles, E. Flat Plate, F. Flexible Tubing,
G. Valve, H. Soap Solution Reservoir.

nitrogen tank and regulator using a constant pressure of
10-20 psig. The size of the bubbles blown in this manner
is fairly uniform and can be varied by changing the orifice
size or the pressure.

With a little experience it is possible to produce
fairly perfect single crystal rafts in the rings which may
then be rotated into desired orientations. It is also
possible to introduce grain boundaries and dislocations
during the filling of the rings or later by stirring.

When the desired array has been produced the valve to
the reservoir is opened and the level of the solution in
the tray is raised to a position several millimeters above
the tops of the forming rings. The free floating arrays of
bubbles exert long range attractive forces upon each other,
analogous to Van der Waals forces, which cause them to move
into contact. Sintering then begins and is usually completed in 15-30 seconds. In this work, the process was recorded on 16 mm. movie film shot at 30 frames per second.

III <u>Results</u>

The spontaneous sintering of various close packed and
square packed arrays of circles and of circle on plate

arrays with bubble sizes ranging from 1 to 3 mm. was
filmed. In all cases the observed sintering of the bubble
raft arrays occurred in a manner analagous to dislocation
shear processes. In some cases certain arrays sintered to
100% density spontaneously. The photographs reproduced
here are enlargements of individual frames of the 16 mm.
movie films. The sequences are chosen to show some of the
general features of the sintering of bubble rafts and are
typical of the rest of the data. The dislocations are
easily seen by viewing the photographs at an angle down the
close packed directions.

The first sequence is the sintering of a circular raft
of 2 mm. diameter bubbles, 10 cm. in diameter, sintering to
a linear edge of a similar crystal containing a surface
ledge at "A" and a grown-in dislocation at "B" on slip "S".
Fig. 2a shows the system as the first contact is formed.

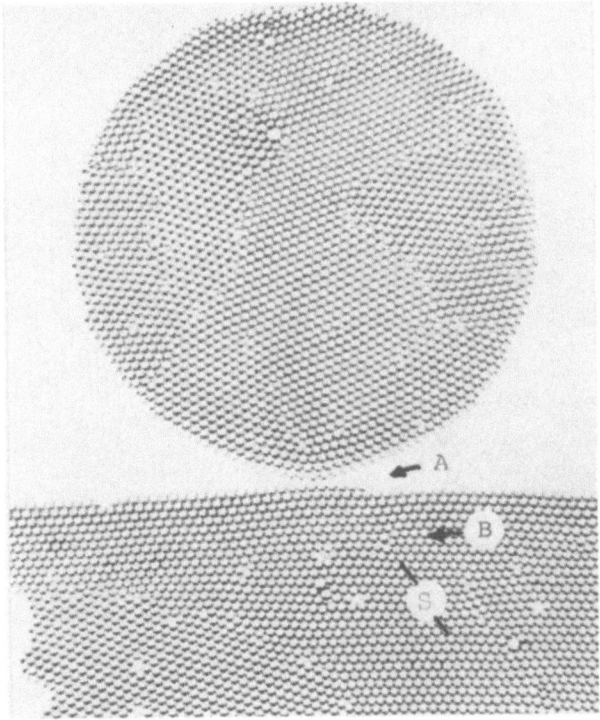

Fig. 2a. Circle on flat plate array, 2 mm. bubbles,
0 sec., A. Atomic Step, B. Grown in Dislocation,
S. Slip plane of B.

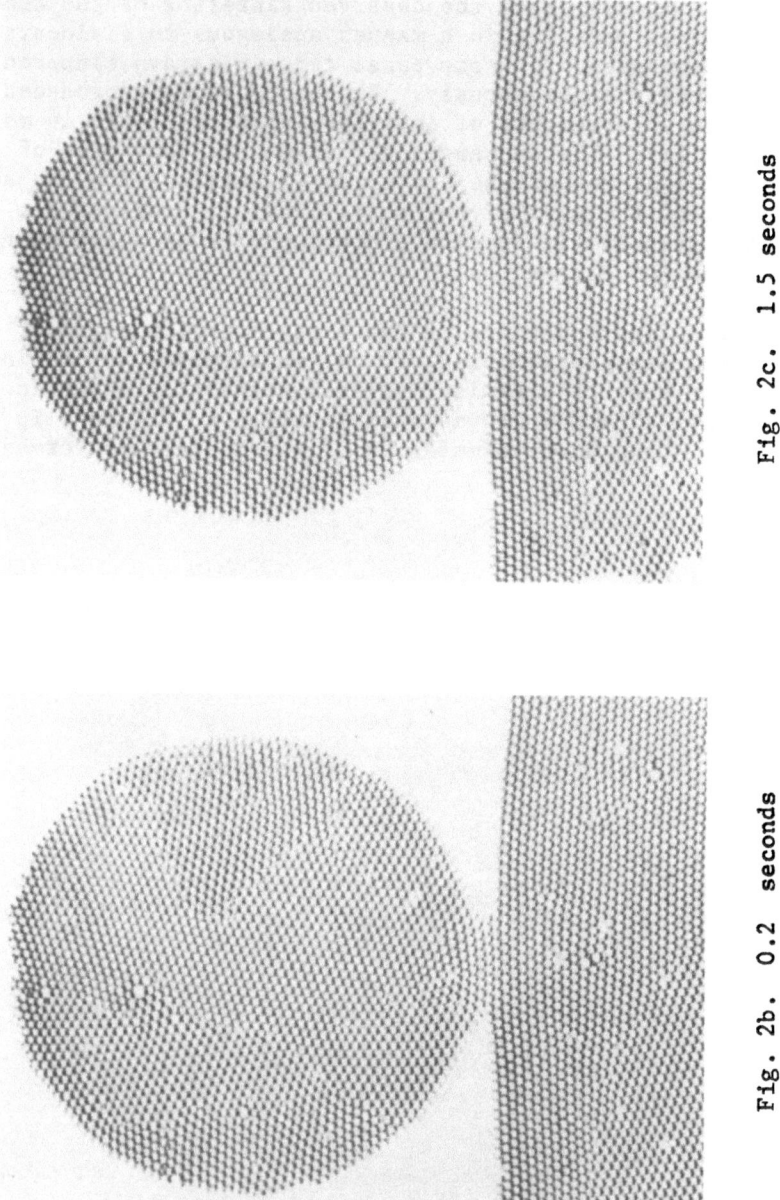

Fig. 2c. 1.5 seconds

Fig. 2b. 0.2 seconds

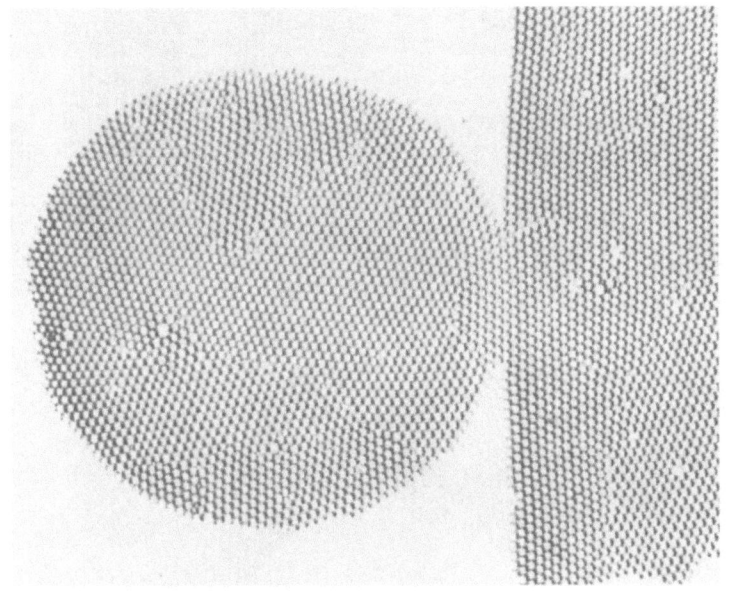

Fig. 2e. 1.7 seconds

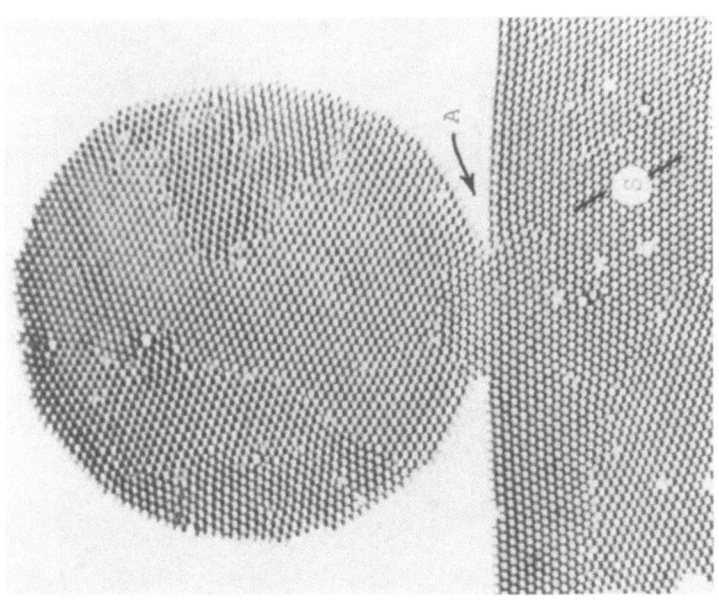

Fig. 2d. 1.6 seconds, A. Dislocation
Nucleation at step, S. Slip plane of
grown in dislocation and new dislocation.

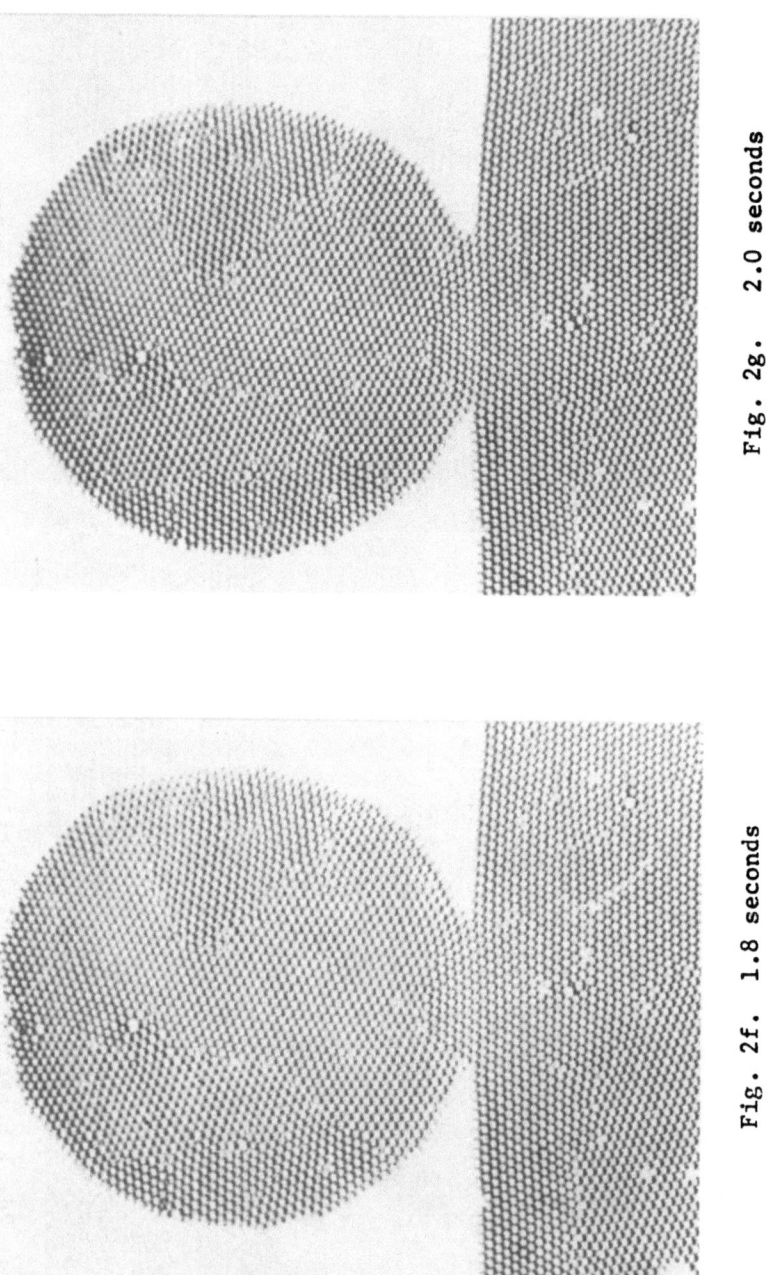

Fig. 2g.   2.0 seconds

Fig. 2f.   1.8 seconds

The neck in Fig. 2b is mainly the result of elastic deformation, the elastic energy having been supplied by the
annihilation of free surface. In Fig.2c slip of grown-in
dislocations close to the neck in the circular particle has
relaxed some of the elastic deformation and allowed more
free surface to be eliminated.

In Figs. 2d, 2e, 2f, and 2g a dislocation is nucleated
at the ledge site "A" in the flat particle along slip plane
"S" glides into the flat particle and combines with the
grown in dislocation at "B". The dislocations annihilate
each other, being of opposite sign and on the same slip
plane. As a result of this event the neck grows and the
particle centers approach each other. The increase in neck
diameter is roughly equal to the component of the Burger's
vector of the nucleated dislocation perpendicular to the
flat surface divided by the tan θ, where θ is the angle between the tangents to the surfaces of the impinging particles at the apex of the neck. The strain energy of the
dislocation is the direct result of the work done by the
forces between approaching surfaces in the vicinity of the
apex of the neck against the image forces which tend to
pull the dislocation out of the crystal.

The sintering system in the second series is a seven
particle close packed hexagonal array of 3.75 cm. diameter
circular particles composed of 1.2 mm. diameter bubbles.
As was mentioned above, this is the bubble size for which
the calculated bonding forces in the bubble raft most
nearly approximate those calculated for copper ions. The
simulated copper particle size is 90 Å. It would be expected that the forces giving rise to dislocation nucleation and slip in this system would be similar to those in
an array of 90 Å diameter Cu particles.

In Fig. 3a the array is shown just after all but one
of the contact points have formed. In Figs. 3b, 3c, and 3d
two dislocations on parallel slip plane at "A" and "B"
nucleate and glide a short distance into the left hand
crystal in a manner quite similar to that shown in Figs. 2d
and 2e. This process is often observed in the sintering of
bubble rafts and will be discussed in more detail in a
following section.

At "C" in Fig. 3d another dislocation nucleates in the
same way and glides into the center crystal in Fig. 3e.

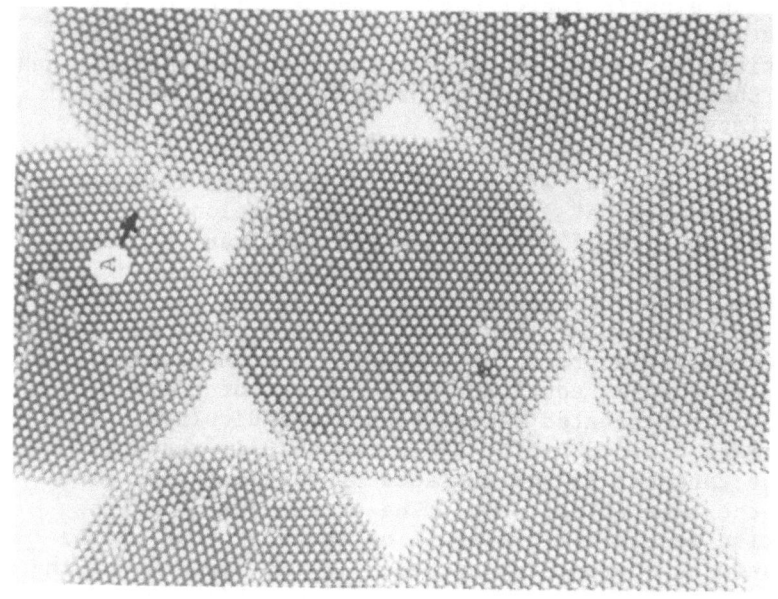

Fig. 3b. 2.2 seconds, A. Dislocation Nucleation at pore corner

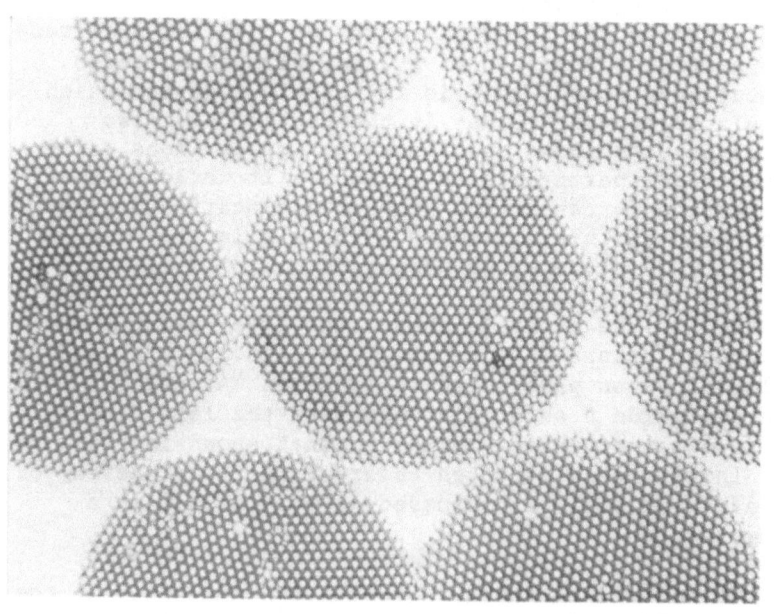

Fig. 3a. Close packed array, 1.2 bubbles, 2 seconds

Fig. 3d. 3.0 seconds, C.
Dislocation Nucleation at
pore corner

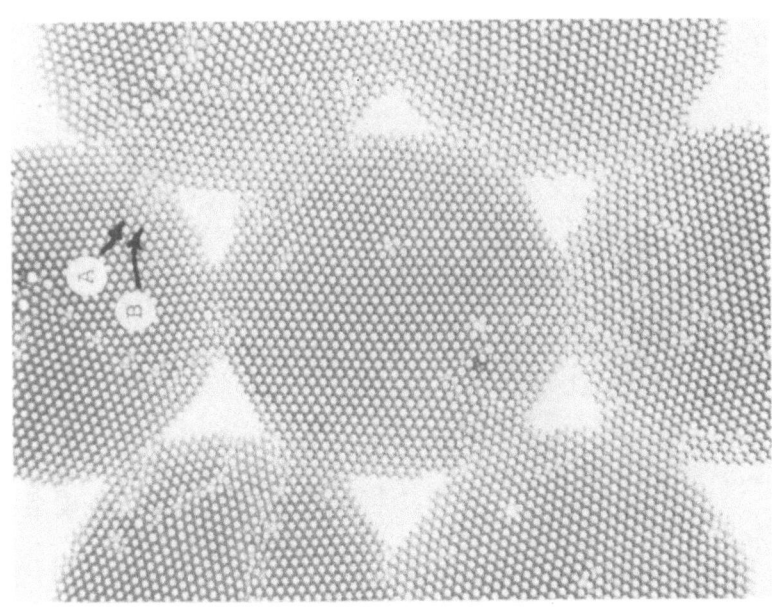

Fig. 3c. 2.6 seconds, A & B
Dislocation Nucleation

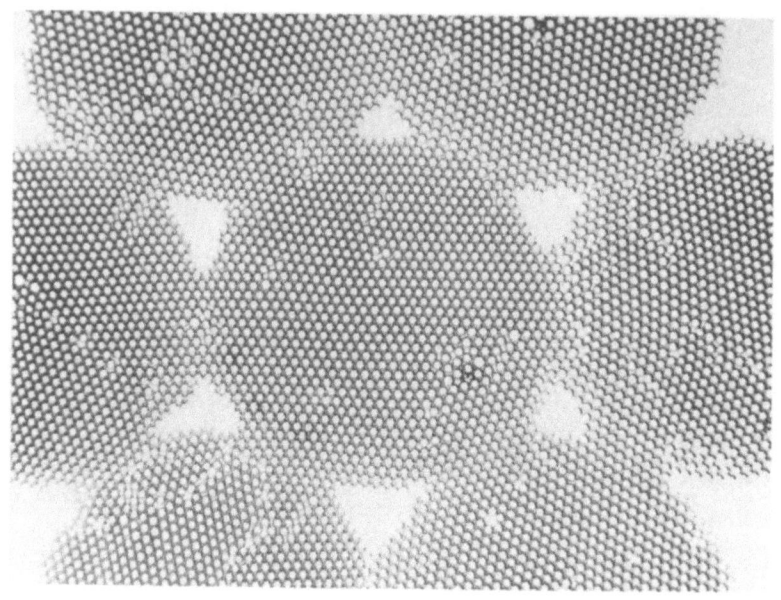

Fig. 3f.   3.4 seconds

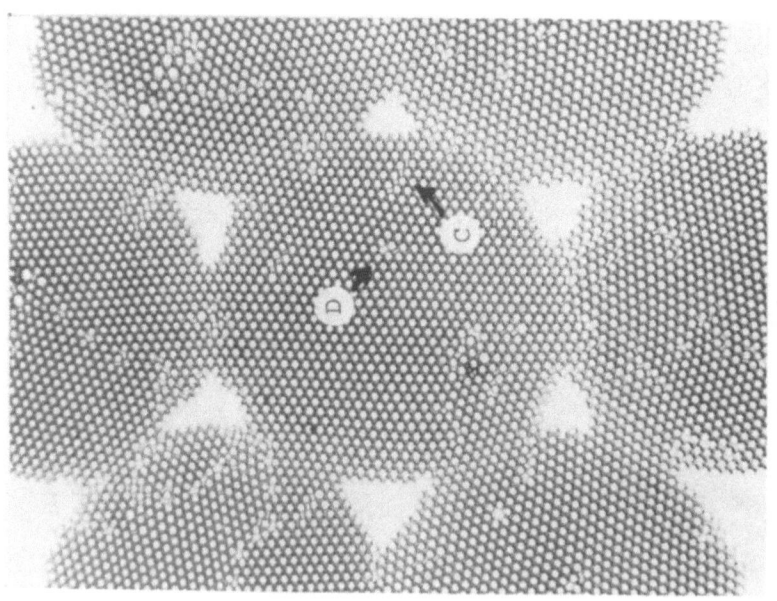

Fig. 3e. 3.2 seconds, C. Gliding
new dislocation. D. Grown in defect

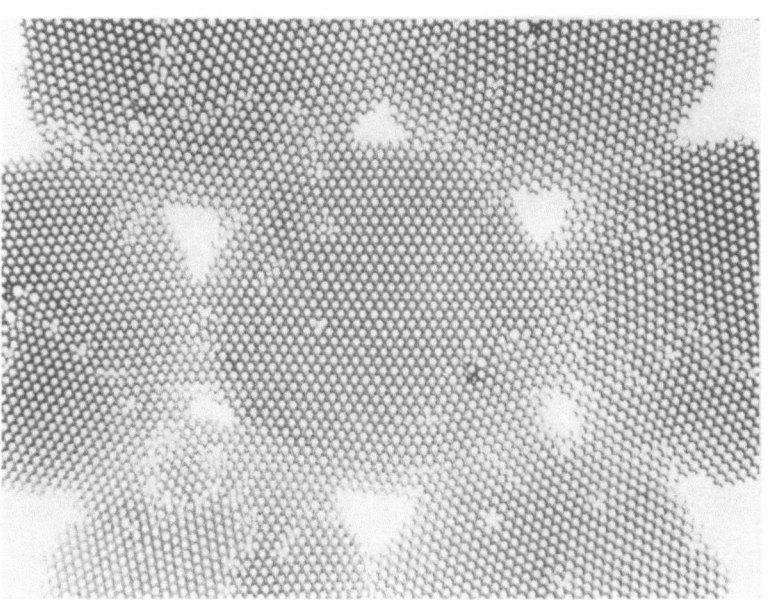

Fig. 3h. 4.6 seconds

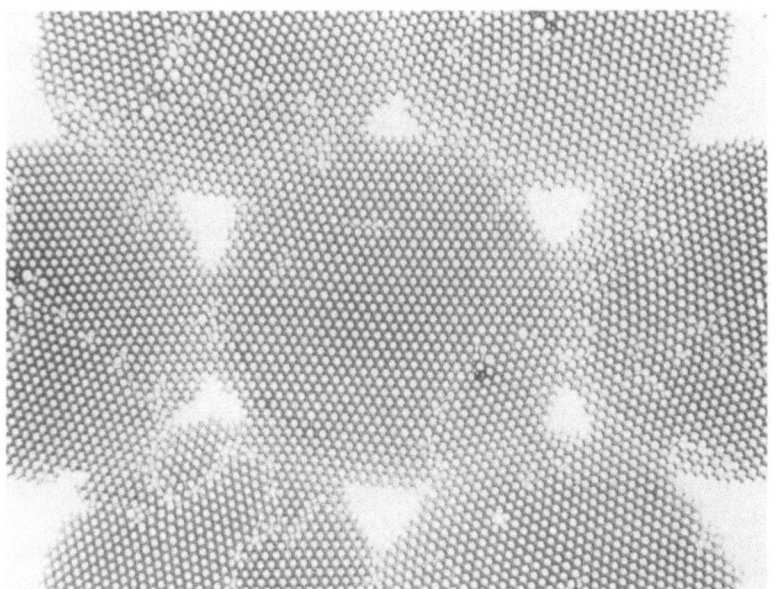

Fig. 3g. 3.8 seconds

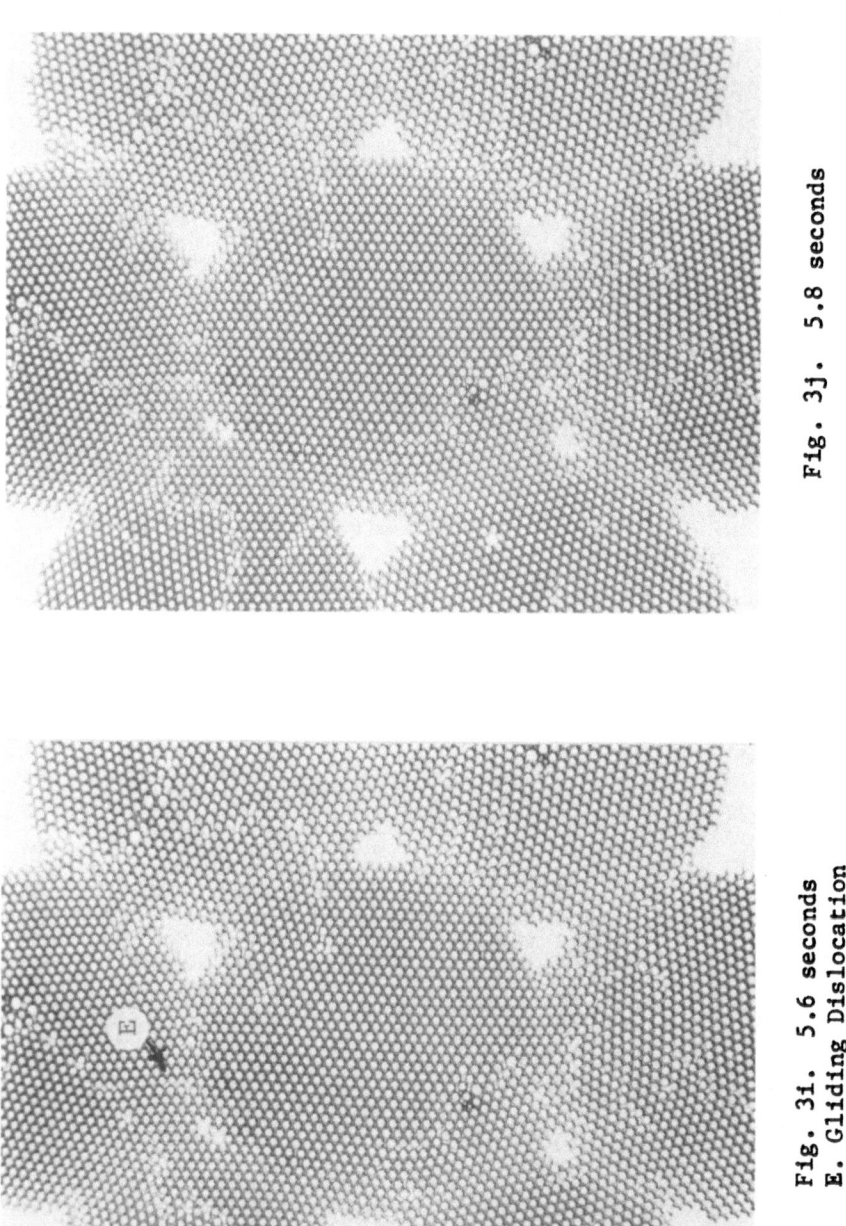

Fig. 3j.	5.8 seconds

Fig. 3i.	5.6 seconds
E. Gliding Dislocation

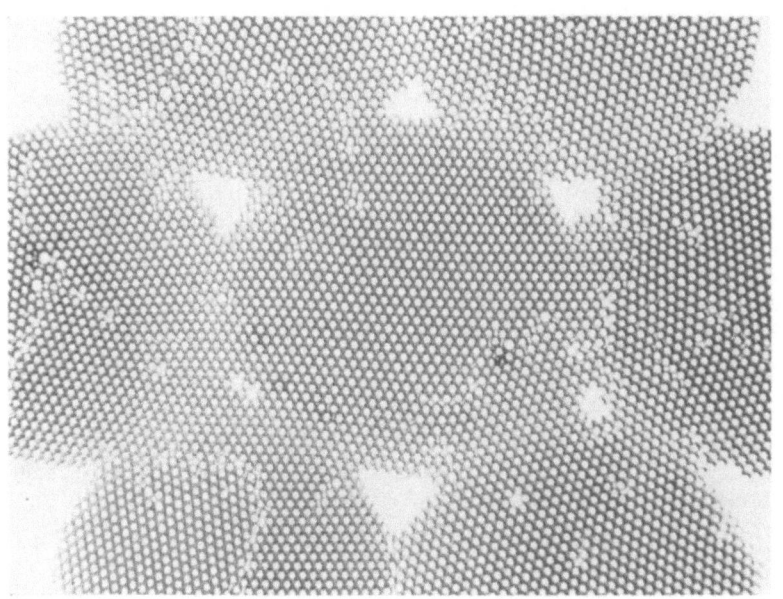

Fig. 31.   6.2 seconds

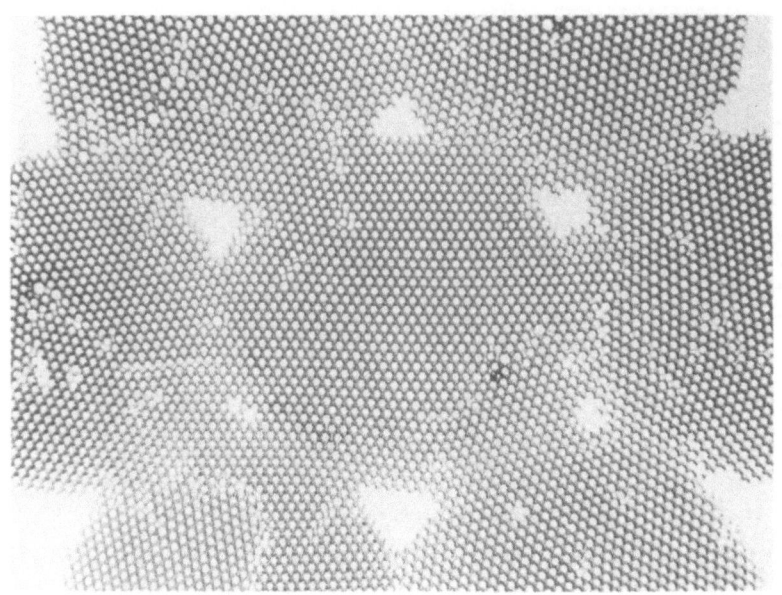

Fig. 3k.   6.0 seconds

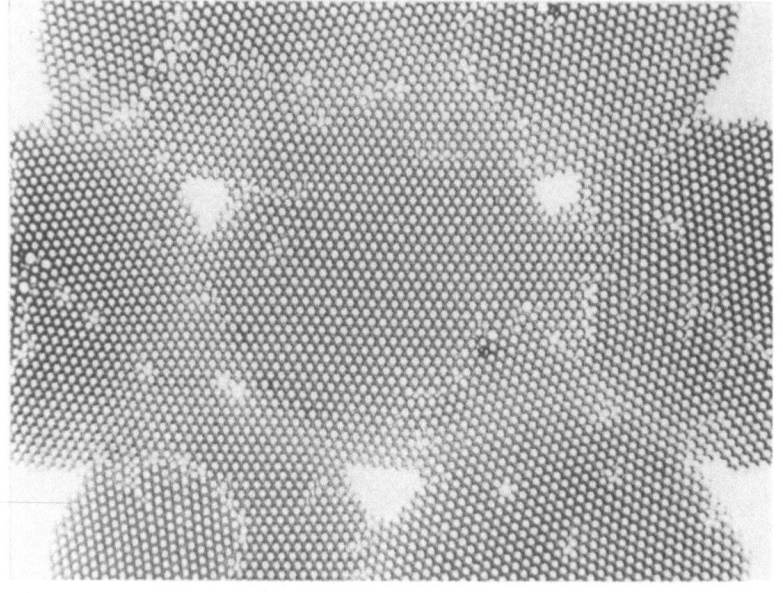

Fig. 3n. 12 seconds

Fig. 3m. 10 seconds

Between Figs. 3f and 3g, it combines with a grown in defect
at "D" and cross slips onto a new slip plane in Fig. 3g.
It stops in Fig. 3h and remains stationary until in Figs.3m
and 3n it glides upward, contributing to the filling of the
two nearby pores.

Notice also the dislocation at "E" in Figs. 3i-3l
which glides upward and undergoes mutual annihilation with
one of the dislocations in the low angle tilt boundary.

Other instances of dislocation motion and grain boundary migration and sliding in response to changing stresses
will be revealed by careful examination of the photographs.

The following general observations were made regarding
the sintering of bubble rafts:

1. The formation of necks, center to center motion of
particles and the shrinkage and elimination of pores in
arrays of bubble rafts occur entirely by elastic and plastic
deformation.

2. The plastic deformation takes place as a result of the
slip of grown-in dislocations, the nucleation and slip of
new dislocations at the pore corners, low angle boundary
migration by slip perpendicular to itself, and shear
sliding parallel to high angle boundaries.

3. Gliding dislocations frequently cross-slip in response
to local variations in the state of stress, a process which
may be a necessary condition for the observed changes in
geometry.

4. The apexes of the necks formed by these processes (the
corners of the pores) do not become rounded as a result of
plastic flow sintering but remain sharp. The angle θ between the adjacent surfaces increases as sintering proceeds.

5. A maximum neck diameter (corresponding to a maximum θ)
is eventually reached beyond which no further slip occurs.
This maximum x/a ratio varies somewhat with the orientation
of the impinging particles and their initial internal grain
boundary configuration and dislocation substructure, and is
relatively insensitive to changes in the particle diameter.

In hexagonal arrays, similar to the one shown here,

the average maximum x/a ratio is about 0.5, the x/a ratio
required for impingement of necks and consequent closure of
pores in a two dimensional close packed array (6 nearest
neighbors). In some cases close packed arrays of 4.7 and 9
particles sintered to 100% density; in other cases one or
more pores remained open indefinitely. In squares packed
arrays, pore closure was never obtained.

6. The shear processes contributing to the growth of one
neck or the shrinkage of one pore often affect and interact
with those occurring at an adjacent neck or pore, that is
to say there are constraint effects due to variations in
the rates of sintering of the various necks in a given
array and contributions from the slip of one dislocation to
shear in various areas of the array.

7. The state of stress at various points in the sintering
array is a function of both the geometry of the surfaces
and the stress fields of the mobile and sessile dislocations present at any instant.

IV. Discussion

The most important source of new dislocations in the
early stages of the sintering of arrays of bubble rafts
seems to be nucleation, at the sharp apexes of the necks of
dislocations in slip directions more or less parallel to
the line joining the centers of the particles (shown in
Figs. 2d, 2c, and 3d). Since the forces between individual
bubbles are known, the forces acting between two parallel
or inclined surfaces separated by some distance could be
calculated by summing over all pairs of particles on
opposite sides of the separation. The force between two
parallel surfaces integrated with respect to distance from
infinite separation to intimate contact would give twice
the surface energy of the crystal.

It is the attractive force between the two adjacent
surfaces which drives the nucleation of these dislocations.
These attractive forces result in a compressive stress in
the neck which is concentrated in the immediate vicinity of
the corner due to the notch type of geometry. The combination of the compressive stress immediately inside the
neck and the tensile stress outside of it generates a large
shear component at the apex, making this point by far the
most favorable nucleation site in the system. As the

dislocation is nucleated and slips into the crystal,
against the image forces tending to draw it toward the
surface, the areas of particle free surface in the vicinity
approach each other and come into contact over a length
given roughly by the component of the Burger's vector of
the dislocation divided by the tangent of the angle θ between the tangents of the surface at the apex. The amount
of energy available to the dislocation nucleation process
is the integral over both surfaces of the force between
them with respect to their relative approach. The amount
of surface energy available for conversion into dislocation
strain energy decreases repidly with increasing θ or x/a
ratio. Thus the existance of a critical x/a ratio or θ
value above which no further nucleation at this site occurs
would be expected. If sintering depended strongly on this
source of dislocation this angle would define the limiting
x/a ratio which could be obtained by slip and this seems to
be the case in arrays of bubble rafts. This type of nucleation site may also be important in the initial contact
formation between sintering metal particles before surface
diffusion has rounded the corner and perhaps more important
in plastically deformable ionic crystals such as CaF where
surface diffusion is less important. It is interesting to
note that before initial contact is made and as long as the
corner remains sharp there is no long range chemical potential gradient on the surfaces driving surface diffusion
due to changes in curvature, although there would be a
potential gradient due to Van der Waals interactions between a surface adatom and ions in the approaching adjacent
free surface.

Calculations of the net free energy change of a dislocation nucleated in this manner between silver spheres
and the free energy change required for the formation of a
critical sized half loop of dislocation indicate that the
reaction is favorable kinetically and thermodynamically
over a wide range of temperatures and θ for atomically
clean surfaces (11).

V. Summary

Films of the sintering of bubble rafts show that the
observed sintering process, from neck growth to pore
closure, can occur entirely by elastic deformation and
dislocation shear processes. An important step in the
observed reactions seems to be dislocation nucleation at

the sharp corners marking the edges of the necks. It is
suggested that the initial sintering of soft metals may
occur by processes similar to those observed in the bubble
rafts since the inter-bubble forces have been shown to
closely resemble those in a metal lattice, and the multiplicity of slip systems in the metal lattices is clearly
sufficient to accomplish the required changes in shape.

References

1. Morgan, C. S. and Yust, C. S., 1963 Journal of Nuclear
   Materials 10, 3, p. 182.

2. Morgan, C. S., McHargue, C. J. and Yust, C. S., 1965
   Proc. Brit. Ceram. Soc. No. 3, p. 177.

3. Early, J. G., Lenel, F. V. and Ansell, G. S., 1964
   Trans. AIME Vol. 230, p. 1641.

4. Hingorany, A. R., Lenel, F. V., and Ansell, G. S., 1969
   Materials Science Research, Vol. IX, Plenum Press.

5. Nunes, J. J., Lenel, F. V., Ansell, G. S., "The
   Influence of Crystalline Anisotropy on Neck Growth
   During the Sintering of Zinc", to be published.

6. Morris, R. C., Lenel, F. V., and Ansell, G. S.,
   "Sintering of Loose Spherical Copper Powder Aggregates
   Using Silica Particles as Markers", to be published.

7. Bragg, W. L., and Nye, J. F., 1947 Proc. Roy. Soc. A
   190, p.474.

8. Bragg, W. L., and Lomer, W. M., 1949 Proc. Roy. Soc. A
   196, p.171.

9. Lomer, W. M., 1949 Proc. Roy. Soc. A 196, p. 182.

10.Sutton, W. H., Modern Composite Materials, Broutman,
   L. J., and Krock, R. H., editors, Addison Wesley,
   Reading, Mass., 1967, p. 414.

11. Morris, R. C., unpublished calculations.

EXPERIMENTAL QUANTITATIVE MICROSCOPY WITH SPECIAL APPLICATIONS TO SINTERING

R. T. DeHoff
E. H. Aigeltinger

Dept. of Metallurgical & Materials Engrg.
University of Florida
Gainesville, Florida 32601

I. INTRODUCTION

Quantitative metallography[1] is a collection of procedures for determining geometric properties of microstructures from observations made upon two-dimensional microsections produced through the structure. Such studies are conveniently classified into three categories, according to whether the observations are made (1) in reflection microscopy; (2) in transmission microscopy; or (3) by serial sectioning* techniques. The present paper avoids the discussion of transmission quantitative microscopy, because this technique has not been applied broadly to sintered structures, and because the use of transmitted light, which forms a projected image, introduces complications into the quantitative analysis of the structure

---

*Serial sectioning is a technique by which a three-dimensional structure is analysed by reconstruction from a collection of closely spaced, parallel microsections.

which require that simplifying geometric assumptions or
approximations be introduced. No such assumptions are
required for the analysis of images obtained in reflection
microscopy, or in serial sectioning.

Accordingly, the paper is divided into two main sections, the first dealing with reflection microscopy and the
second with serial sectioning. Each of these major sections is further divided into the following topics:

1. A list and description of the three dimensional
   geometric properties which can be determined;

2. A list and description of the observations that
   are required to determine the three dimensional
   properties;

3. Presentation of some examples of each kind of
   measurement;

4. A presentation of the statistics of each kind of
   measurement; and

5. Principles for design of experiments.

These sections are followed by a discussion which con-
trasts the view of sintering provided by quantitative
microstructural analysis with that provided by other
approaches to studying the process.

II. QUANTITATIVE REFLECTION MICROSCOPY

Figure 1 shows the evolution of structure during sintering of an uncompacted single size fraction of spherical
copper powder, as observed in reflection microscopy.
Each of the micrographs shown is obtained by producing
an essentially plane section through the three-dimensional
structure and photographing this polished section in vertical illumination. A list of the structural features that may
exist in a single solid phase sintered system is presented

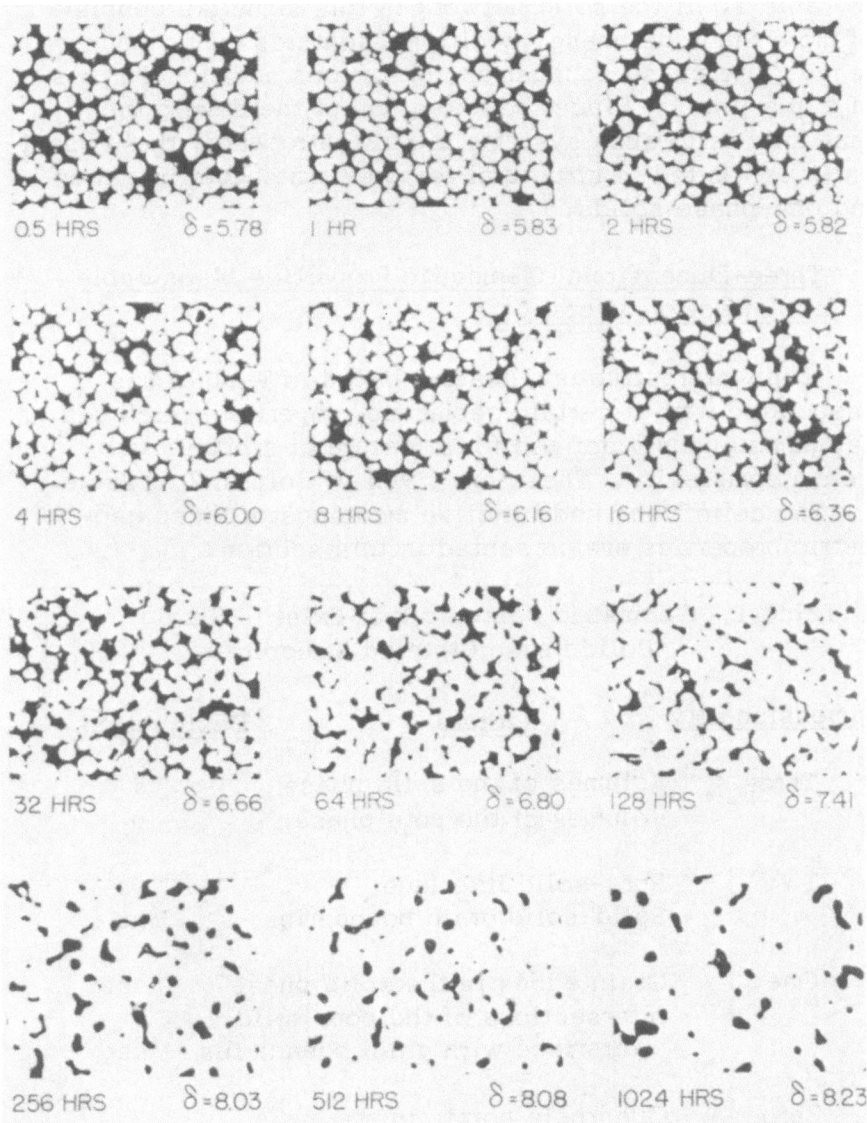

Fig. 1. Evolution of microstructure during sintering of a (-170 +200) mesh size fraction of spherical atomized copper powder sintered without compaction in dry hydrogen at 1000 ± 5°C.

in Table 1. If the solid network in this structure consisted
of more than one phase, additional features would appear
(e.g., interphase boundaries, or various additional kinds
of triple lines). Since the extension of the development
that follows to such systems is straightforward, the present
paper is limited to structural features that occur in sintering one-phase solids.

A. Three-Dimensional Geometric Properties Measurable
   in Reflection Microscopy

Each microstructural feature listed in Table 1 has
associated with it certain geometric properties which can
be unambiguously defined for a feature of arbitrary geometric complexity. These are given in Column 2 of Table
2. The definitions and intuitive meanings of these geometric properties are presented in this section.

Table 1. Geometric Features that Exist in Single
Solid Phase Sintered Structures

Dimensionality	Feature	Designation
Three	Volumes of the solid phase	s
	Volumes of the pore phase	p
Two	Pore-solid interface	ps
	Solid-solid grain boundaries	ss
One	Grain edges in the solid phase	sss
	Intersections of the pore solid interface with grain boundaries	ssp
Zero	Quadruple points in the grain boundary network	ssss
	Intersections of three grain boundaries on the pore-solid interface	sssp

Table 2. Properties Associated with Geometric Features Occurring in Sintered Structures

Feature	Property	Symbol	Units
Volume of Solid	Volume Fraction	V_V^s	cm^3/cm^3
	Mean Solid Intercept	$\bar{\lambda}_s$	cm
Volume of Pores	Volume Fraction	V_V^p	cm^3/cm^3
	Mean Pore Intercept	$\bar{\lambda}_p$	cm
Pore-Solid Interface	Surface Area	S_V^{ps}	cm^2/cm^3
	Total Curvature	M_V^{ps}	cm/cm^3
	Mean Surface Curvature	$\bar{H}$	$1/cm$
	Genus	G_V	$1/cm^3$
	Number of Separate Parts	N_V	$1/cm^3$
Grain Boundary	Surface Area	S_V^{ss}	cm^2/cm^3
	Mean Grain Intercept	$\bar{\lambda}_g$	cm
Grain Edges	Total Edge Length	L_V^{sss}	cm/cm^3
Triple Lines on Pore-Solid Interface	Total Line Length	L_V^{ssp}	cm/cm^3

1. <u>Volume Fraction</u>, V_V. The volume fraction of the
pore phase is the total volume of porosity contained in unit
volume of microstructure. The corresponding property may
be defined for the solid phase, and, in this two-phase
system clearly,

$$V_V^s + V_V^p = 1 \qquad\qquad (1)$$

The volume fraction is the most straightforward measure
of the <u>amount</u> of porosity (or solid) present in the system.

2. <u>Surface Area</u>, S_V. Each kind of interface in the
system (i.e., the pore-solid interface, and the grain
boundaries in the solid phase) has a surface area associated with it; S_V is the total surface area of a particular
kind of interface contained in unit volume of microstructure.
S_V values for various kinds of interface are not related.

3. <u>Length of Triple Line</u>, L_V. According to Table 1,
two types of triple lines may exist in a sintered structure;
grain edges (type sss) and the intersection of grain boundaries with the pore-solid interface (type ssp). Each kind
of lineal feature has a length associated with it. L_V is the
total length of a particular lineal feature contained in unit
volume of structure.

4. <u>Total Curvature</u>, M_V. Interphase interfaces, or,
more generally, those interfaces in a system which have a
distinguishable inside and outside,* possess a property
called the <u>total curvature</u>. The curvature at a point on a
surface is defined in terms of two quantities, called the
<u>principal normal curvatures</u>, k_1 and k_2. These quantities
are obtained by erecting a surface normal, and intersecting the surface with all possible planes that contain this
normal. Each such intersection is a curve; the curvature
of this arc at the point on the surface is defined in the usual
way for a two-dimensional curve, i.e., as the reciprocal
of the radius of a circle which passes through three adjacent points on the curve. As the intersecting plane is
rotated about the surface normal, the curvature of the arc

_____

*This quantity cannot be evaluated for grain boundaries.

of intersection varies between limits. The limiting values of curvature, i.e., the maximum and minimum curvature produced by intersecting planes containing the normal, are defined as the <u>principal normal curvatures</u> for the surface at the point in question.

The mean curvature, H, is defined as the average of the principal normal curvatures:

$$H \equiv \frac{1}{2} (k_1 + k_2) \qquad (2)$$

The Gaussian curvature is defined as their product:

$$K \equiv k_1 k_2 \qquad (3)$$

In general, the principal normal curvatures, and consequently the mean and gaussian curvatures, vary from point to point on a surface. The <u>total curvature</u> of a surface (or a collection of surfaces) is obtained by multiplying the mean curvature H in an element of surface by the area of the element ds , and integrating the result over the area of the surface:

$$M \equiv \iint_S Hds \qquad (4)$$

The total curvature per unit volume of structure M_V is the total curvature of all of the surface of a particular kind contained in some volume of the structure, divided by the volume.

5. <u>The Average Mean Surface Curvature, $\overline{H}$</u>. Since the mean curvature varies from point to point on a surface, it has some distribution of values, and therefore some average value. This may be defined as:

$$\overline{H} \equiv \frac{M_V}{S_V} \qquad (5)$$

$\overline{H}$ may be positive or negative (depending upon which phase is taken as inside the surface, and which outside); large values of $\overline{H}$ correspond to sharply curved surfaces, and values near to zero to relatively flat surfaces. Since the chemical potential near a curved surface, and the normal pressure acting upon a curved surface from surface tension,

is proportional to the local value of the mean curvature[2],
this quantity may be a measure of the average driving force
in sintering under the action of surface tension.

6. <u>Mean Intercept of a Phase</u>, $\bar{\lambda}$. A straight line
drawn through a sintered structure will form intercepts with
the solid and pore phase volumes. For a given phase, these
intercepts will exhibit a distribution of lengths and will
thus possess an average value, $\bar{\lambda}$. This quantity may be
defined for phase volumes of arbitrary shape, size, or distribution, and is thus a measure of the <u>scale</u> of the system.
It can be shown that: [3]

$$\bar{\lambda}_p = \frac{4V_V^p}{S_V^{ps}} \tag{6a}$$

$$\bar{\lambda}_s = \frac{4V_V^s}{S_V^{ps}} \tag{6b}$$

where S_V^{ps} is the surface area of the pore-solid interface.

An analogous quantity is the mean grain intercept, $\bar{\lambda}_g$,
which is defined as the average surface to surface distance
of grains in the structure. In this case, the surface area
of the particles in question becomes $S_V^{ps} + 2S_V^{ss}$, since
each grain boundary is counted twice. The corresponding
relation for mean grain intercept is then:

$$\bar{\lambda}_g = \frac{4V_V^s}{S_V^{ps} + 2S_V^{ss}} \tag{6c}$$

B. <u>Geometric Properties Which Can Be Measured on the
    Sectioning Plane</u>.

The four counting measurements described in this section do not comprise a complete list of observations that may
be made upon a representative microsection through a

sintered structure. However, these measurements share the unique features that (a) they are very simple and straight-forward observations to make, and (b) they are ambiguously related to the three-dimensional geometric properties defined in the previous section.

Consider the two-dimensional microsection of a par-tially sintered copper powder aggregate shown in Figure 2. A grid of lines has been superimposed upon this micro-structure. All four of the measurements described in this may be made with the aid of this grid. In practice, the grid may be superimposed by placing a clear plastic overlay with the grid drawn on it over a photomicrograph, or by inserting a graticule with an inscribed grid in the eyepiece

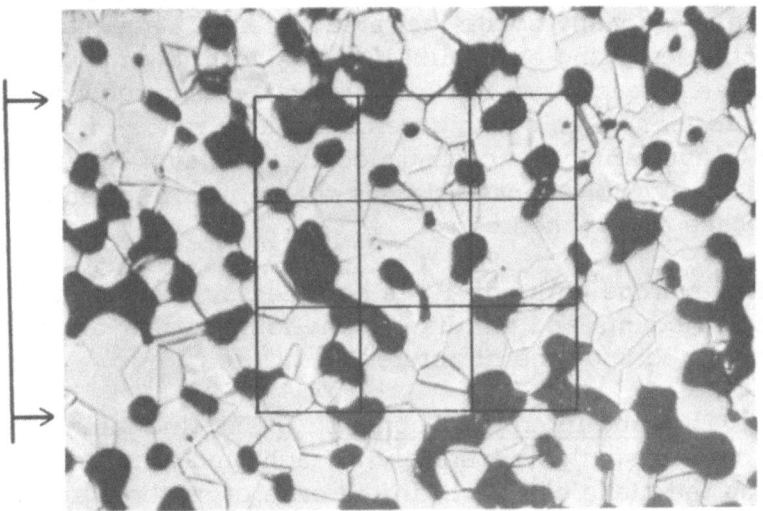

Fig. 2.   Microstructure of −200+230 mesh spherical Cu powder sintered to a pore volume fraction of 0.17.   The superim-posed grid is used for quantitative microscopic measurements. The vertical line at the left may be swept over the area to make the tangent count.   Magnification about 120X.

of the microscope. The latter procedure is generally to be preferred, because (a) it avoids the necessity for producing photographs; (b) it facilitates a uniform sampling of the polished area of the specimen, to help insure representativeness; and (c) by permitting direct observation of the specimen under the microscope, this procedure retains the capability to adjust the focus for observations that require careful decisions.

1. <u>The Point Count, P_p</u>. The point count may be applied to either phase in the system; for the following discussion, suppose that it is being applied to the pore phase. The number of points in the grid which lie in the pore phase is counted and tabulated. The point count for the pore phase P_p is the ratio of this number to the total number of points in the grid. For the structure shown in Figure 2, this number is 0.31. In order to optimize the procedure for most ranges of porosity, it is most convenient to use either a nine point (3 X 3) or a 16 point (4 X 4) grid. More dense grids give too many counts per area observed, and make subjective counting errors likely; grids with fewer points consume too much time in moving the stage and recording the observations. If an easily interchangeable range of grid sizes is available, the optimum conditions favor the use of progressively more dense grids as the volume fraction óf pore phase approaches zero.

The average value of P_p is obtained by summing the individual counts and dividing by the total number of points superimposed upon the structure.

2. <u>The Line Intercept Count, N_L</u>. The line intercept count may be applied separately to each kind of interface trace on the plane of polish; in the case of sintered structures, it can be applied to the pore-solid interface traces, and to the grain boundary surface traces. To illustrate the count, consider the traces of the pore-solid interface shown in Figure 2. Select one or more of the lines in the superimposed grid. These test lines are scanned, and the number of times they cross traces of the pore-solid interface is

counted. The real length (actual length as superimposed on the structure) of these lines must be separately determined with a stage micrometer. The line intercept count N_L^{ps} for a single placement of the grid is the ratio of the number of intersections with pore-solid interface to the actual length of test line scanned. The analogous count can be made for grain boundary traces.

The average value of N_L may be obtained by summing the counts on individual areas, and dividing by the total length of test line scanned in all of the observations. For representative results, the microsection should be scanned with a uniform distribution of placements of the grid, and a uniform distribution of orientations of the grid.*

3. <u>The Area Tangent Count, T_{Anet}</u>. The area tangent count is applied to traces of interfaces in the structure which have an inside and an outside. In sintered structures, this limits this count to traces of the pore-solid interface. An area delineated by one or more of the squares in the super-imposed grid is chosen for scanning. In general the area should be selected so that a total of 5 to 15 counts is typically obtained in each observation. The scanning area must be measured, using a stage micrometer. The tangent count requires that this area be swept over by a cross hair. A Filar eyepiece with a grid graticule may be used for this purpose. The number of times this sweeping test line forms tangents with elements of the pore-solid interface trace is counted and tabulated. It is necessary to separately tabu-late tangents with convex segments of arc, and tangents with concave arc; consequently, it is necessary to establish a convention as to which side of the surface is "inside" and which is "outside". Let us choose the solid phase to be the inside of the surface. Then a tangent is counted as

---

*If the structural features involved are isotropically distri-buted in orientation, a single orientation of the test lines will provide a representative sample.

positive if the arc it touches is bent towards the solid phase
(convex); a tangent is counted as negative if the arc it
touches is bent towards the solid phase (convex); a tangent
is counted as negative if the arc it touches is bent towards
the pore phase (concave). The net tangent count is the
difference between the numbers of convex and concave
tangents counted in the area. The area tangent count, T_{Anet},
is then defined as the ratio of this difference in counts to
the area scanned by the moving cross hair.

The average value of T_{Anet} for the structure is obtained
by separately summing the number of positive and negative
tangent counts, obtaining the required difference, and dividing by the total area scanned in the set of observations on
the structure.

4. The Triple Point Count, P_A. Lineal features in the
three-dimensional structure appear as points on a metallographic sectioning plane. The lineal features that exist in
sintered structures are triple lines of the type (sss) (grain
edges) and (ssp) (intersection of a grain boundary with the
pore-solid interface), see Table 1. These appear as corresponding triple points on the microsection. The triple point
count may be applied separately to each of these kinds of
triple points. An area is selected for the count; generally
this may be one or more square sections of the grid. This
area is then measured. Since the triple point count involves
a feature-to-feature scan, which is difficult to perform
accurately under the microscope, the size of the grid should
be chosen so that only a few counts will be obtained in each
grid square. The number of triple points of the type being
determined is then counted. The triple point is the ratio of
the number of triple points thus counted to the area scanned
during the count.

The average value of P_A is obtained by summing the
counts for each observation, and dividing by the total actual
area scanned.

The literature of quantitative microscopy makes it clear

that there are alternative measurements that can be made that are essentially equivalent to the four kinds of measurements outlined in this section. For example the familiar lineal analysis[4,5], in which a test line is superimposed upon the structure, and the fraction of the line length lying within the phase of interest is measured, is equivalent to the point count. Similarly, the measurement of the total length of pore-solid interface traces per unit area of structure is equivalent to the line intercept count. These measurements may be useful if automatic scanning equipment is available[6,7]. However, the techniques described in this section are to be preferred for manual quantitative microscopy because: (a) they are all simple counting measurements, and do not require measurements of lengths and areas; (b) they are all counts of points, thus minimizing errors that might be associated with counting objects that cross the boundary of the area of observation; and (c) they can be made directly under the microscope with no equipment more complex then a Filar eyepiece with a grid graticule.

C. The Fundamental Relationships of Quantitative Microscopy.

The quantities described in the preceding section derive their essential utility from the fact that they provide unbiased estimates of the geometric properties of the three-dimensional structure they represent. The relationships between the microsection observations and the three-dimensional geometry constitute the fundamental relationships of quantitative microscopy. These equations are based upon a combination of integral geometry and probability theory, and are derived elsewhere[1,8,9]. In order for them to be valid, it is only necessary that the sample observed be representative; specifically this means that any microstructural gradients or anisotropies be uniformly represented in the sample. It is also possible to obtain quantitative estimates of the degree of anisotropy, and the magnitude of gradients that may exist[1], but this is beyond the scope of the present paper.

The fundamental relations are:

$$V_V = \overline{P}_P \tag{7}$$

$$S_V = 2\overline{N}_L \tag{8}$$

$$M_V = \pi\overline{T}_{Anet} \tag{9}$$

$$L_V = 2\overline{P}_A \tag{10}$$

where all of the quantities involved have been defined in the preceding two sections. The derived quantities expressed in equations (5) and (6) may also be expressed in terms of these counting measurements:

$$\overline{H} = \frac{\pi}{2} \frac{T_A^{ps}}{N_L^{ps}} \tag{11}$$

$$\overline{\lambda}_p = 2 \frac{P_P^p}{N_L^{ps}} \tag{12a}$$

$$\overline{\lambda}_s = 2 \frac{P_P^s}{N_L^{ps}} \tag{12b}$$

$$\overline{\lambda}_g = 2 \frac{P_P^s}{N_L^{ps} + 2N_L^{ss}} \tag{12c}$$

It is thus clear that a great deal of detailed geometric information about the microstructure may be obtained from these simple counting measurements; specifically, all of the metric properties listed in Table 2 may be so determined.

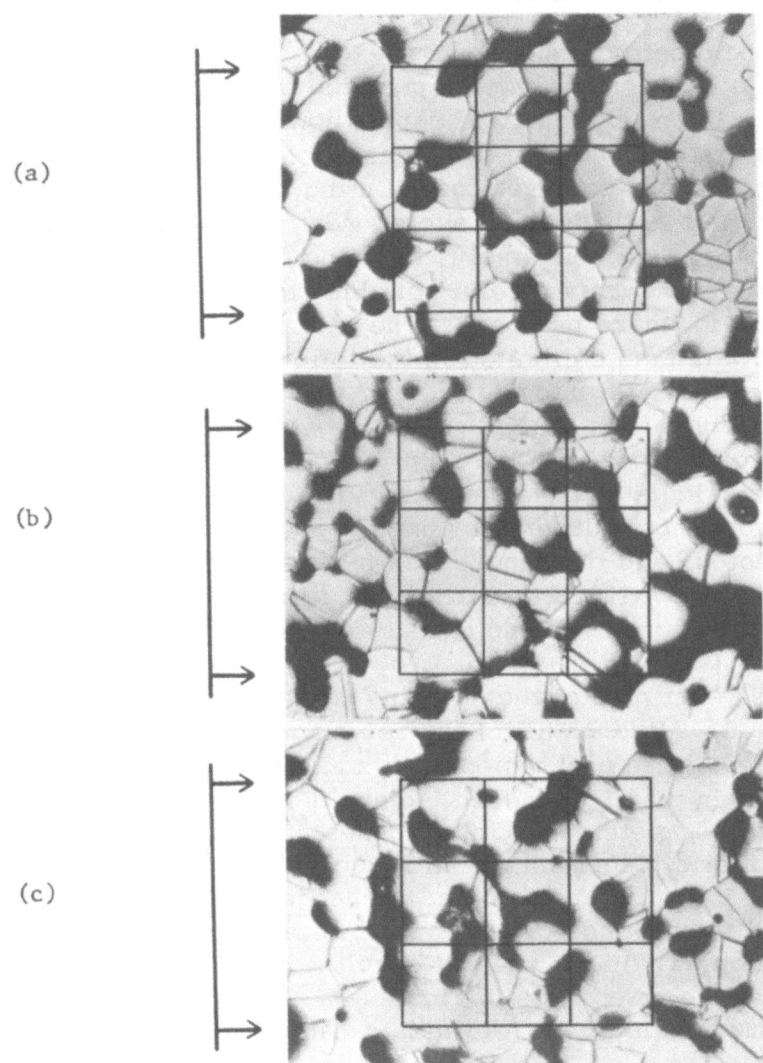

Fig. 3.   Three areas taken from the same specimen of -200
+230 mesh spherical copper powder sintered in dry hydrogen
at 1010°C to a pore volume fraction of 0.17.   Counts made
with the superimposed grids are tabulated in Table 3.
Magnification about 120X.

D. Examples of Application of the Counting Measurements.

The microstructures shown in Figure 3 are different
areas of a single microsection prepared from a sintered mass
of -200 +230 mesh spherical copper powder, sintered for
128 hours in dry hydrogen to a density of 7.41 gm/cc, as
determined by water immersion. During metallographic
preparation it is difficult to avoid smearing the copper over
the pore area, or, conversely, rounding the edges of the
copper phase. A very careful polishing technique was
developed for these purposes. The polished section thus
prepared was then calibrated in the following way. The
point count was applied to the structure, with 25 areas
being observed with a 3 X 5 grid. The volume fraction of
porosity calculated from the point count was then compared
with that calculated from the density determination. If
these two numbers differed by more than 2.5%, the polish
was rejected, and the sample was taken back to 3/0 paper
for repolishing. If it was within these limits, the polish was
accepted, and the counting measurements were made.

Three examples of the counting measurements are shown
in Figure 3. Although all of the counting measurements are
made on the same grid region in this example, this is not
necessary in practice; indeed, to avoid confusion, it is
helpful to complete each kind of observation for the required
number of grid placements before proceeding to the next. In
addition, it is often convenient to change the magnification
for the different kinds of observations, or to use different
grid graticules. In the present case the point count was
applied to the pore phase in Figures 3a, b, and c; the
results are tabulated in Column 2 of Table 3.

The line intercept count was obtained by scanning the
horizontal lines of the grids in these three figures. Counts
were made separately for the pore-solid interface, and for
the grain boundaries in the structure; these are tabulated in
Columns 3 and 4 in Table 3. Although both counts are performed on the structures as shown, in practice it may be
helpful to make the count of the pore-solid trace intercepts

Table 3. Counting Measurements from Figure 3.

Micrograph	P_P	N_L^{ps}	COUNTS N_L^{gb}	T_A^{ps*}	P_A^{sss*}	P_S^{ssp*}
a	2	19	12	-7 +4	13	16
b	3	15	12	-9 +3	12	23
c	3	18	12	-12 +1	12	24
Sum	8	52	36	-28 +8	37	63
Dimension Scanned	--	0.12 cm	0.12 cm	3×10^{-4} cm^2	9×10^{-4} cm^2	9×10^{-4} cm^2
Total	--	0.36 cm	0.36 cm	9×10^{-4} cm^2	27×10^{4} cm^2	27×10^{-4} cm^2
Normalized	--	144.3 cm^{-1}	100 cm^{-1}	-2.22×10^{4} cm^{-2}	1.37×10^{4} cm^{-2}	2.33×10^{4} cm^{-2}
Geometric Properties	V_V 0.167	S_V^{ps} 288.6 $\frac{cm^2}{cm^3}$	S_V^{ss} 200 $\frac{cm^2}{cm^3}$	M_V^{ps} -6.97×10^{4} $\frac{cm^3}{cm}$	L_V^{sss} 2.74×10^{4} $\frac{cm^3}{cm}$	L_V^{sss} 4.66×10^{4} $\frac{cm^3}{cm}$
Calculated Properties	$\bar{\lambda}_p$ 23.1×10^{-4} cm		$\bar{\lambda}_s$ 115×10^{-4} cm	$\bar{\lambda}_g$ 48.4×10^{-4} cm	$\bar{H}$ -2.41×10^{2} cm^{-1}	

*Counted top three grid areas.

on an as-polished surface; the count of grain boundary trace intersections must of course be made upon an etched section.

The area tangent count was obtained for the top three grid squares on each micrograph, by sweeping the vertical line indicated in Figure 3 across the structure in the horizontal direction. The positive and negative tangent counts are given in Column 5 of Table 3.

The triple point count, which also must be made upon an etched section, was applied separately to the grain edges (sss) and the triple points on the pore-solid interface traces (ssp). These counts were also applied to the top three grid squares. The results are tabulated in Columns 6 and 7 of Table 3.

The dimensions of the grid, determined from measurements made with a stage micrometer, are 0.03 X 0.03 cm. The scanned regions (lineal and areal) for each count are given in Row 5 of Table 3. The total regions scanned are shown in Row 6. The normalized values of the counting measurements are then obtained by taking the ratio of corresponding values in Row 4 and Row 6, and are given in Row 7. Equations (7) through (10) were then applied to these quantities to give the resulting estimates of the three-dimensional properties of the structure. In addition, mean intercepts of porosity, solid, and grains were calculated using equations (6). The average value of the mean surface curvature for this structure is negative, indicating that the pore-solid surface is more concave than convex. These results are compiled at the bottom of Table 3.

The information given in Table 3 provides a very detailed (though not complete, in itself) quantitative description of the three-dimensional microstructure under investigation. Additional properties that may be determined are topological, and will be discussed under the subdivision on serial sectioning.

Of course, it is not necessary to measure all of these

quantities for specific studies. In much of the studies on
the evolution of microstructure during sintering carried out
at the University of Florida, for example, attention was
limited to the pore-solid interface. Thus the quantities
studied were the volume fraction of the pore phase, and the
surface area and total curvature of the pore-solid interface.
The grain structure was investigated in only a limited number of cases.

E. Statistics of the Counting Measurements.

A long range study of the evolution of microstructure
during sintering conducted at the University of Florida has
provided a wealth of statistical information about many of
the counting measurements. Quantitative microscopy has
been carried out on structures that cover the full range of
microstructural conditions, from 0.05 to 0.90 volume fraction of porosity, and in the size range from 10 to 100 microns. These data have been analysed for their representative average values to gain insight into the sintering
process[10]. They have also been analysed for the corresponding values of standard deviation, to provide insight
into the statistics of the counting measurements in quantitative microscopy. These statistical results form the
basis for the design of experiments in quantitative microscopy.

1. Statistical Background. In the design of experiments in quantitative microscopy, the most important deci-
sion centers around how many measurements must be made.
This is closely related to the level of accuracy sought for
the description of the microstructure. A single kind of observation may be expected to produce a range of counts as
the grid is moved about over the microstructure. This distribution of results has a mean value, $\bar{x}$, which is the
average number of counts for the set of observations, and
some dispersion of values, which is best described by the
standard deviation s_x for the set of readings, and the
coefficient of variance, C.V. If a set of n counts is made
(a sample of size n), the standard deviation of the
counted quantity, x, is given by (11):

$$s_x = \frac{\sum\limits_{i=1}^{n} (x_i - \bar{x})^2}{n - 1} \qquad (13)$$

where the x_i values are the counts obtained for each individual observation.

The coefficient of variance is defined by:

$$C.V. \equiv \frac{s_x}{\bar{x}} \qquad (14)$$

and therefore expresses the dispersion as a fraction of the mean value.

If additional samples of size n were taken from the structure, and the mean values of these samples calculated, it would in general be found that the sample means also cover a range of values, although this range is considerably smaller than that observed for individual readings. The central limit theorem of statistics[11] states that the sample means are normally distributed, with mean value $\bar{x}$ equal to the mean of the population μ_x and a standard deviation given by:

$$\sigma_{\bar{x}} = \frac{\sigma_x}{\sqrt{n}} \qquad (15)$$

where σ_x is the standard deviation of the population distribution. The measured standard deviation s_x is used to estimate σ_x. Thus, the dispersion of the mean values of samples of size n decreases with the square root of the number of readings in the sample. This leads to the statistical rule of thumb that the accuracy of an estimate increases with the square of the effort expended.

Because the sample means are normally distributed, it is possible to assert that for an indefinitely large collection of samples, 95% of the resulting sample mean values will lie within $\pm 2 \sigma_{\bar{x}}$ of the mean value of the variable for the

structure. Conversely, the probability that the mean of a
given sample of size n will lie within $\pm 2\sigma_{\bar{x}}$ of the mean
of the structure is 0.95. The range of values corresponding
to $\pm 2\sigma_{\bar{x}}$ is called the "95% confidence interval" for the
measured sample mean, and the ends of this interval are
called the 95% confidence limits.

For example, suppose a total of 100 placements of the
grid are made, and the line intercept count is obtained for
each placement. The average for this sample of 100 observations is calculated to be 28.7 counts and the standard
deviation, employing equation (13) is 5.1 counts. From
equation (15), the standard deviation of the mean of the
sample is $5.1/\sqrt{100} = 0.51$; accordingly, the confidence
interval is ± 1.02 counts, and the 95% confidence limits
are 27.68 and 29.72. This may be interpreted as meaning
that the true mean value for the line intercept count for the
structure lies between 27.68 and 29.72 with a probability
of 0.95. If the total length of test line in these 100 samples
is 0.105 cm, the estimate of N_L for the structure is
273 ± 9 (1/cm). Applying equation (8) gives the corresponding estimate for the surface area: $S_V = 546 \pm 18$ cm^2/cm^3.

The coefficient of variance defined in equation (14)
gives the same kind of information, expressed in the percent
of the average value. Thus, for the above example, the
coefficient of variance per observation is 0.178; the coefficient of variance of the mean values of 100 observations is
obtained by dividing by $\sqrt{100} = 10$ or 0.0178; and the confidence interval for the number of counts, and the estimates
of N_L and S_V are all the same when expressed in percent
of the variable; $\pm 1.78\%$. These percentages are an estimate of the statistical accuracy of the result, for the observations taken.

The design of experiments, on the other hand, implies
that a level of accuracy, usually expressed as a percent of
the mean value, is preselected, and the basic question
becomes, "How many counts must be made to achieve
this level of accuracy?" Suppose a level of accuracy of y

percent is chosen as acceptable for the desired result. This
implies that the confidence interval $(2\sigma_{\bar{x}})$ is equal to y
percent of the mean value, or, mathematically

$$2\sigma_{\bar{x}} = (\frac{y}{100})\bar{x} = 2\frac{\sigma_x}{\sqrt{n}}$$

Solve for the number of readings required:

$$n = (\frac{200}{y}\frac{\sigma_x}{\bar{x}})^2 = (\frac{200}{y} C.V.)^2 \quad (16)$$

Thus, for a given microstructure, one might begin by taking
a small number of observations, and estimating the mean and
standard deviation. These results are then substituted into
equation (16) to calculate n, and the experiment carried
to completion. The full set of n readings is then analysed
to determine the estimate of the mean and the confidence
limits. This result may be compared to the accuracy initially specified, and adjustments made if necessary.

F. The Design of Experiments in Quantitative Reflection
   Microscopy.

Sophisticated attempts to provide theoretical estimates
of the coefficient of variance of the point count have met
with limited success[12]. The results obtained are valid
for random distributions of the particles of the phase of
interest, and are thus limited to small volume fractions.
Similar attempts to predict the variance of a set of counts
for N_L, T_{Anet}, and P_A have not appeared in the literature.

However, the extensive program on quantitative microscopy of sintered structures carried out at the University
of Florida has provided the basic for making empirical estimates of the coefficient of variance of three of the counting
measurements; (a) the point count; (b) the line intercept
count; and (c) the tangent count. These empirical correla-
tions are summarized in this section and used as a basis

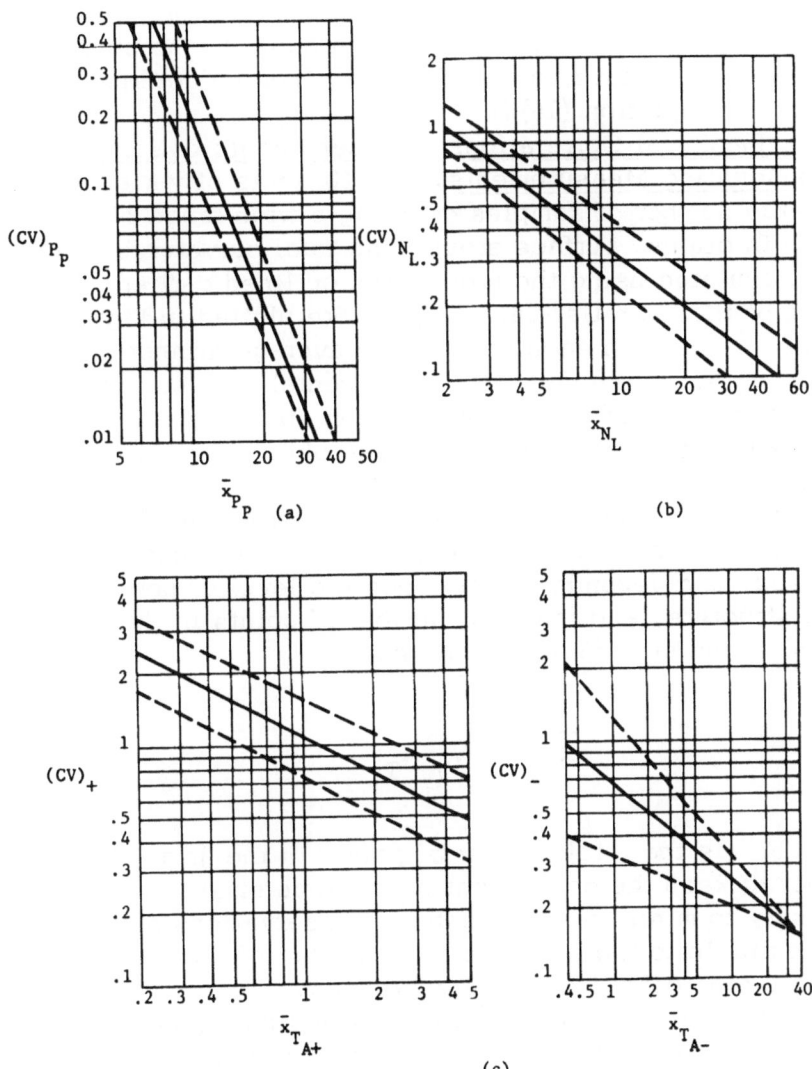

Fig. 4. Empirical correlation between the coefficient of variance of the point count, line intercept count and tangent count and the number of counts for each placement of the grid.

for the design of experiments. Once the coefficient of variance has been estimated for a particular structure, and a
reasonable level of accuracy has been chosen, the number
of observations that must be made may then be determined
by applying equation (16).

    1. <u>Empirical Correlations for Coefficients of Variance
of the Counting Measurements</u>. Coefficients of variance
of a large number of samples covering a wide range of pore
volume fractions, surface areas, and total curvatures, have
been calculated using the procedures outlined earlier. The
results of these statistical analyses are plotted in Figure
4. The abscissa in each case is the average number of
counts (points, intersections, or tangents) obtained for a
single placement of the grid on a given structure. The
ordinate is the coefficient of variance obtained for the same
experiment.

    In the point count, Figure 4a, the average number of
counts in an observation is independent of the magnification used, and depends chiefly upon the level of porosity
in the structure, and the total number of points in the grid
applied to the structure. In this case

$$\bar{x}_{P_P} = V_V \, N_T \qquad\qquad (17)$$

where N_T is the total number of points in the grid. Thus,
in order to keep the coefficient of variance small, which is
necessary to minimize the number of observations required,
it is desirable to increase the total number of points in the
grid as the porosity level (V_V) decreases.

    The number of counts obtained for a single placement
of the grid for the line intercept count Figure 4b depends
upon the level of porosity, and the scale of the microstructure (i.e., average size), and the magnification used, as
well as the dimensions of the grid. If the scale of the
system is characterized by the mean pore intercept, $\bar{\lambda}_p$
(see equation (6a)), the average number of intersections
expected with a total length of grid lines at the image plane,
L_G, is

$$\bar{x}_{N_L} = 2 \ \frac{V_V \ L_G}{M \bar{\lambda}_p} \tag{18}$$

where M is the magnification. Note that the number of intersections increases with porosity and length of grid line counted per placement of the grid, and decreases with increasing magnification and scale of the structure. For a given structure, V_V and $\bar{\lambda}_p$ are fixed, so that control of the value of $\bar{x}_{N_L}$ may be exercised by selecting the magnification and the number of grid lines scanned for each placement. Large values of $\bar{x}_{N_L}$ decrease the coefficient of variance, and reduce the total number of grid placements necessary to achieve a given level of accuracy. The selection of magnification and grid spacing must be limited by the requirement that the total number of counts per observation should not exceed 20 or 30; otherwise subjective counting errors will be introduced.

The average number of counts obtained in the determination of the area tangent count for a single placement of the grid depends essentially upon the same parameters as does the N_L count. If A_G is the area within the grid, measured at the image plane, which is selected to be analysed, then the average number of tangents formed within this area is roughly

$$\bar{x}_{T_A} = \frac{4k}{\pi} \ \frac{A_G \ V_V}{M^2 \bar{\lambda}_p^2} \tag{19}$$

where k is an unknown constant. The number of counts increases with the actual grid area observed in each placement, and with the level of porosity; it decreases as the scale of the system and the magnification are increased. For a given structure, for which V_V and $\bar{\lambda}_p$ are fixed, control of $\bar{x}_{T_A}$ is accomplished by selection of the magnification, and by choosing the number of squares in the grid which will be scanned for the count. The tangent count is facilitated by choosing a rectangular area, rather than a square one. Since the tangent count involves scanning a length of line as it sweeps across an area, it is easier to scan a short line segment, and move the segment over a relatively long distance.

2. <u>Rules of Thumb in Designing Quantitative Metallographic Experiments</u>. In order of increasing difficulty, the counting measurements are: (1) the point count; (2) the line intercept count; (3) the area tangent count; and (4) the triple point count. This ordering derives from the fact that the unit experiments involve respectively: (1) a point-to-point scan on a grid; (2) a feature-to-feature scan along a line; (3) a feature-to-feature scan along a line as it moves over an area; and (4) a feature-to-feature scan over an area. The number of counts that can reliably be made on a single observation decreases as the difficulty of the measurement increases. For manual counts made under the microscope this number should range from 20 to 30 counts for the easy observations to about 5 to 15 for the difficult ones.

It is generally desirable that the grid spacing be at least of the order of the scale of the structure (i.e., the spacing of features), and preferably somewhat larger.

It is better to sample the structure lightly with a large number of observations, rather than with a few grid placements, each with a large count. This procedure has the dual advantage of reducing the likelihood of subjective counting errors, and assuring that the total sample will be representative. The individual placements should be uniformly distributed over the structure. This procedure will give representative results so long as the structure is not periodic, with a period simply related to the spacing between observations.

In any of the counting measurements, a number of placements of the grid which give a total of about 500 counts will yield an accuracy of the order of 3 to 10% in the estimate of the microstructural variable under study. Thus, in a typical experiment, about 10 counts will be made per placement of the grid, and the microscope stage will be moved a total of 50 times. The resulting accuracy will be about 5% of the mean value determined from these measurements.

III. MICROSTRUCTURAL ANALYSIS BY SERIAL SECTIONING

This section is devoted to the description of an analytical technique which allows the determination of the topological properties of the void-solid interface of sintered structures.* The third dimension of a structure must be observed in order to determine the values of these properties. This is presently accomplished by means of a set of closely spaced, parallel microsections called serial sections.

To obtain a set of serial sections, a flat surface is first prepared which passes through the structure and allows the observation of features of interest. This surface is recorded photographically. A thin layer of the structure, parallel to the first surface, is then removed and a new surface is prepared and photographed. This process is repeated until the desired number of serial sections has been obtained.

A. Topological Properties Measurable from Serial Sections

Three topological properties are important in the description of the microstructure of sinter bodies: (1) the genus of the void-solid interface; (2) the number of separate parts of the void-solid interface; and (3) the connectivity of the void or solid space. These topological properties, can also be employed to characterize structures other than those associated with sintered materials. For example the topological properties of grain boundary structures[13,14] and of multiphase materials[15] have been considered.

---

*Structural features are said to be topologically equivalent if they can be made to superimpose by continuous deformation. For example the surface of a coffee cup can be continuously deformed until it assumes the shape of a torus or donut. Topological properties are those which are unchanged by continuous deformation.

Figure 5 shows a sphere with a handle attached. The surface of this sphere with one handle is "topologically equivalent" to the surface of a torus. This means that the sphere with one handle can be continuously deformed until it assumes the shape of the torus.

The genus, or "handle number", G , of a closed sur-face in three-dimensional space can be thought of as the number of handles of the surface. Since a torus is topologi-cally equivalent to a sphere with one handle, the surface of a torus has a genus of one. This concept can be extended to closed surfaces of greater complexity. A closed surface with a genus of G is topologically equivalent to the sur-face of a sphere with G handles attached[16,17].

When the exterior of a sinter body is taken as part of the void space the void-solid interface has a value of genus equal to the number of handles of the interface. That is to say the void-solid interface can be continuously deform-ed into a sphere with handles attached, the number of handles being equal to the genus of the void-solid inter-face.

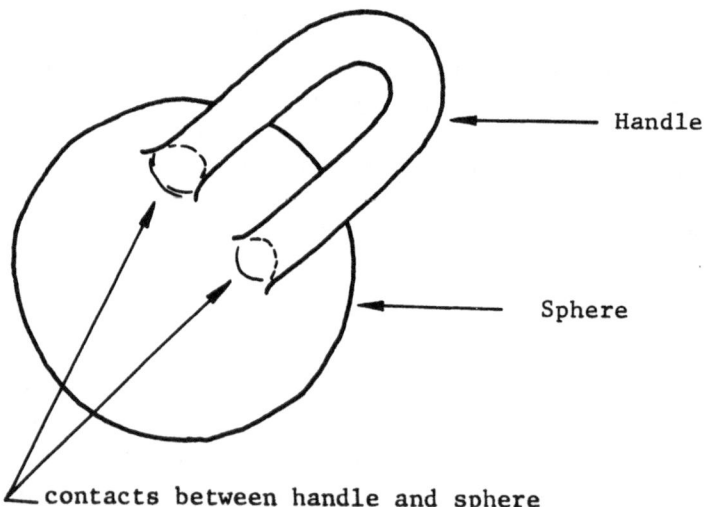

Handle

Sphere

contacts between handle and sphere

Fig. 5. Illustration of a sphere with one handle attached. This closed surface has a genus of one.

The number N of separate parts of the void space of a sintered body, is the number of regions of the void space which cannot be connected to any other region by any continuous path lying entirely within the void space. The interface associated with a separate part of the void space is referred to as a separate part of the void-solid interface. Separate parts of the solid space are not experimentally observed.

The connectivity of the void or solid space is defined as the number of independent, closed paths, or circuits, which cannot be shrunk continuously to a point within the void or solid space[14]. The set of paths or curves are independent if no two can be deformed so that they superimpose without leaving the space, void or material.

The connectivity of a space or volume is equal to the genus of the bounding surface of the space[16,17]. This is apparent, as only paths which encircle "handles" of the bounding surface cannot be shrunk continuously to a point within the void or material space[14]. Since the same surface bounds both the void and material spaces of a sinter body the connectivity of the void and material spaces is the same[16].

An important concept, which will be employed in the following discussion, is that of the deformation retract[18]. The retract of a closed surface is formed, in principle, by shrinking the space bounded by the surface until a network of nodes and branches is formed. For example the retract of the surface shown in Figure 6 is indicated by the dotted line. The dimensionality of the surface is decreased to that of a curve by this process; however, the connectivity and number of separate parts of the surface from which it is obtained. The procedure for the determination of the genus and number of separate parts of the void-solid interface of a sinter structure, presented in the following sections, is based on the determination of the connectivity and number of separate parts of the deformation

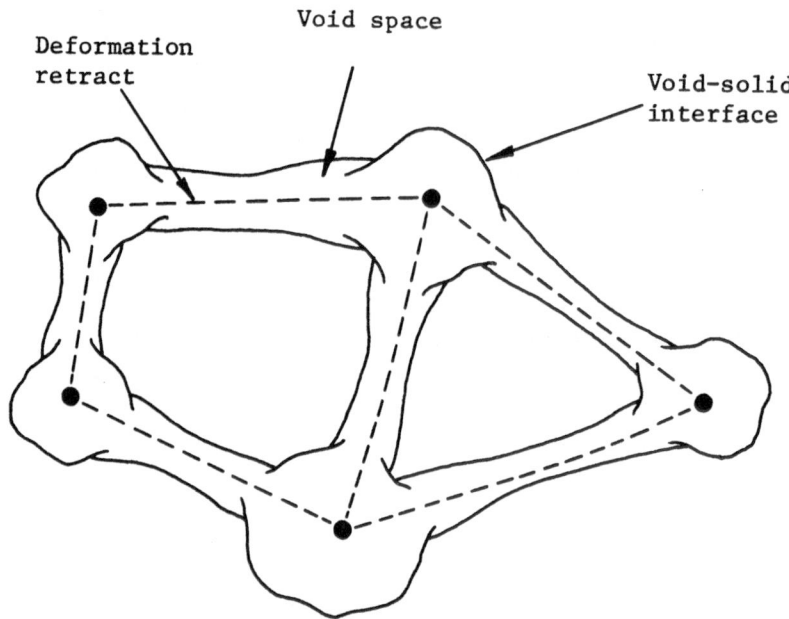

Fig. 6.   Illustration of the construction of the deformation
retract node-branch network corresponding to a multiply
connected closed surface.

retract of the void-solid interface. This network representation of the structure has been called the topological model of sintered systems.

B. Quantities Measured in a Series of Serial Sections

Consider a set of serial sections through a structure
which is multiply connected and possesses separate parts.
An observation area is constructed on each section (usually
square or circular) such that each successive area lies
directly below the preceeding one. In this way a sample
of the structure is contained between the top and the bottom
serial sections and the cylindrical sides established by the

perimeters of the areas on each section.

The surface of the volume established by an arbitrary number of serial sections through the structure in general intersects a number of branches of the deformation retract that represents the structure. These branches must be properly accounted for in determining the topological parameters because the sample analysed is small relative to the entire sinter body. This sample-surface-effect is taken into account by estimating maximum and minimum values that the genus may attain.

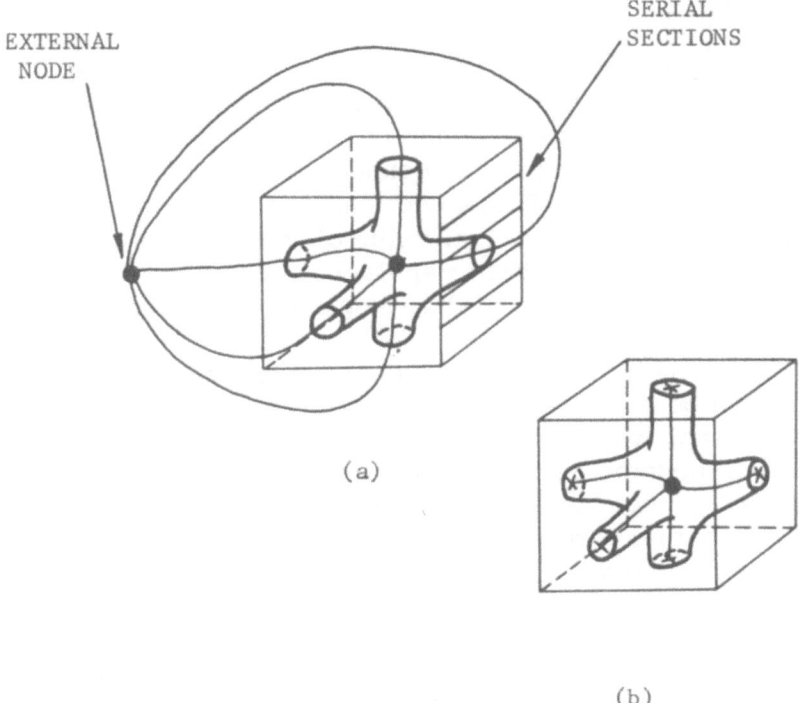

(a)

(b)

Fig. 7. Construction of the limiting networks for a porous body, (a) with all branches crossing the sample surface connected to an external node, and (b) with all branches capped at the surface.

A maximum limit of the genus, G_{max}, is established by assuming that all branches of the retract which cross the surface of the volume analysed are incident upon an "external node", Figure 7a. If X branches are cut by the sample surface, the assumption that they are incident on the external node contributes (X-1) to this estimate of the connectivity of the retract associated with the volume analysed. (Two branches must be incident on the external node before a circuit can be established.) Thus every branch incident on the external node, other than the first branch, contributes one to the genus of the structure contained in the volume analysed. This is the <u>maximum</u> possible contribution of each branch which crosses the surface of the sample.

A <u>minimum</u> limit of the estimate of the genus, G_{min}, is established when it is assumed that all branches of the retract which cross the surface of the sample are capped at the surface, Figure 7b. This construction establishes a lower limit to the value of genus because the capped branches contribute nothing to the connectivity of the retract of the volume analysed. These two limits of the genus, G_{max}, and G_{min}, bracket the actual value of the genus of that part of the sinter body that is analysed. They approach each other as the volume of the sample is increased.

The number of separate parts, N, contained in the sample can be established as follows. Each separate part in the volume, which does not intersect the surface of the volume, is counted as one. Separate parts intersected by the sides, but not the edges of the sample are counted as 1/2 and separate parts intersected by the edges of the sample are counted as 1/4.

During the analysis of a structure each serial section is considered in succession. The size of the "sample" therefore increases as each serial section is added to those already considered.

As a result, when the (i + 1)st serial section is added to the ith serial section, G_{max}, G_{min} and N of the void-solid interface contained in the sample will, in general, increase. The changes in these parameters, ΔG_{max}, ΔG_{min} and ΔN, as successive serial sections are considered, are the quantities which are determined as the serial sections are analysed.

C. Relation Between Measured Quantities and the Desired Parameters.

A general approach for measuring the number of physically identifiable features per unit volume using serial sections, developed by Steele[14], is presented in this section.

In order for this technique to produce a representative value of the number of features per unit volume it is necessary to assume the existence of a unit volume of homogeniety in systems analysed[19]. In general this volume, V_S, is much smaller than the total volume of the structure, V, but larger than the incremental volume between adjacent serial sections. The expected number of features within a sample volume, V_S, is then

$$<N_S> = \overline{N_V} V_S, \tag{20}$$

where $\overline{N_V}$ is the average volume density (number per unit volume) of the features of interest. The topological properties must be estimated on the basis of unit volume or unit mass so that the topological properties of different structures can be compared.

The cumulative number of features observed with the sectioning technique after n incremental volumes have been considered is given by

$$N(n) = N_o + \sum_{i=o}^{n} \Delta N(i), \tag{21}$$

where N_o is the number observed on the initial section and $N(i)$ is the change in the number of features in the ith incremental volume. Thus for larger values of n (i.e., where the cumulative sample volume, $V(n)$, is greater than V_s), the summation in equation (21) can be approximated by $\overline{N_V} V(n)$:

$$N(n) \cong N_o + \overline{N_V} V(n) \qquad (22)$$

The cumulative number of features will therefore increase linearly with volume sectioned, for n sufficiently large. The slope of a plot of the cumulative function, $N(n)$, vs. cumulative sample volume, $V(n)$, is the average volume density of features, N_V. Therefore it is only necessary to determine the change of the cumulative function with cumulative sample volume in order to determine the average volume density of features, N_V. The cumulative volume analysed may be determined by measuring the area analysed on the sections, and the spacing between sections. This spacing can be measured, in most cases, to an acceptable degree of accuracy with a micrometer.

D. Experimental Procedure.

Procedures for unambiguously determining and tabulating changes in the topological properties as a set of serial sections is analysed will now be presented. After the desired area on each serial section is delineated, the next step is the construction of a network in the inscribed area. Either the void or the material space of any sinter structure can be analysed. In general the space with the lowest volume fraction yields the most accurate results[20]. The intersection of the space chosen for analysis with the plane of each serial section is a set of areas bounded by closed curves which are the intersection of the plane of the serial section with the void-solid interface, e.g., see Figure 1. If each of these areas is shrunk continuously in the plane of observation; a set of networks of nodes and branches is formed. This construction is a two dimensional analogue of the deformation retract process in three dimensions.

An example of such a network is shown for a section through
a sinter structure in Figure 8. The network in two dimensions obtained by this procedure will, in general, be composed of a number of isolated networks or separate parts,

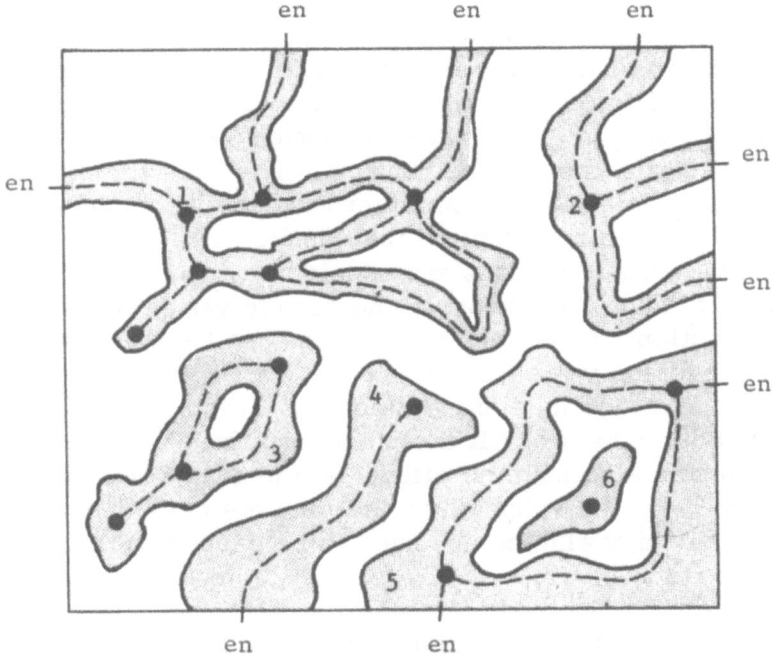

Fig. 8.   Construction of the subnetworks on a schematic
microsection of a sintered structure. This network construction is the two-dimensional analogue of the deformation
retract in three dimensions.

called subnetworks. The retract shown in Figure 8 consists of six subnetworks. These isolated subnetworks may
be connected in the third dimension.

The next, and most important, step in the analysis is
the comparison in succession of the networks on adjacent
serial sections. As the comparison of each successive
network with the preceeding network is made, changes are

noted and tabulated. Four types of changes can occur. The tabulation of these changes is facilitated by means of a labeling procedure, which is described as each type of change is presented.

Throughout the analysis all branches that cross the top and bottom of the volume analysed are assumed to be capped at the surface of the volume. This does not affect the estimate obtained for G_{Vmax} as it is only necessary to determine the change in G_{max} which accompanies the addition of each serial section. The number of branches through the top of the sample remains constant throughout the analysis and therefore has no effect on the change in G_{max} as successive serial sections are considered. The number of branches through the bottom of the sample varies from section to section. However, this number can be characterized by some average value, since it does not vary systematically as successive sections are considered if the structure is homogeneous. As a result the branches through the bottom of the sample also have no effect (on the average) on the change in G_{max} as successive sections are considered. Hence, they may also be assumed to be capped at the sample surface.

The four classes of changes which must be monitored and tabulated are listed in Table 4. Each of these classes will now be described in detail. No attempt is made here to explain why the following procedure allows the determination of the genus and number of separate parts. An explanation of the basis for the procedure is presented elsewhere[20].

Class I: Changes in the Number of Subnetworks. Each subnetwork on the first section is labeled with a different number as shown in Figure 8. As each successive section is considered, corresponding subnetworks are identified and labeled with the same number as on the preceding section. When a new subnetwork appears on a section, (other than the first) and ends on a subsequent section without contacting the surface of the volume analysed or another

Table 4. Classification of Changes Occurring in Serial Sectioning

Class	Description of Change	ΔG_{max}	ΔG_{min}	ΔN	Comment
I	appearance of subnetwork	0	0	0	
	disappearance of subnetwork	0	0	+1	(if subnetwork is not connected to the external node)
II	appearance of new circuit	0	0	0	
	appearance of new branch	+1	+1	0	
	disappearance of a circuit	-1	-1	0	
		0	0	+1	(if the circuit which disappears was the last circuit associated with the appearance of a new circuit)
	disappearance of a branch	0	0	0	
III	separation of a subnetwork into two subnetworks	0	0	0	
	appearance of a connection between two subnetworks	0	0	0	(if labels are different or if one or both subnetworks are new subnetworks)
		+1	+1	0	(if labels are the same)
IV	appearance of a branch to the external node	+1	0	0	
	disappearance of a branch to the external node	0	0	0	
	collapse (or disappearance) of a circuit incident on the external node	-1	0	0	
	appearance of a circuit incident on the external node	0	0	0	

subnetwork, then the void-solid interface associated with
this subnetwork constitutes a separate part. This can only
occur during the analysis of the void space, since separate
parts of the material space are not observed. As a result
the analysis of the void and material spaces is slightly different.

When a new subnetwork appears during the analysis of
the void space it is labeled with a symbol, rather than a
number, the same symbol being used for all new subnetworks.
This allows new subnetworks to be followed from section to
section, just as in the case of subnetworks already present.
The number of separate parts per serial section, N, is determined during the analysis of each serial section, by noting
each separate part when it disappears. (This procedure does
not allow the detection of separate parts which intersect the
sample surface. However, it has been shown experimentally that neglecting these separate parts introduces a small
error into the results.)

Since separate parts of the material space are not observed it is unnecessary to label new subnetworks during
the analysis of the material space. However, so that every
subnetwork on a section will contain some symbol, it is
recommended that new subnetworks be labeled in the analysis of the solid phase, to avoid confusion.

Class II: Changes in the Connectivity of Subnetworks.
Changes in connectivity of each subnetwork must be noted
as successive serial sections are considered. The connectivity of a subnetwork is increased in two ways: by the
appearance of a new circuit, or by the appearance of a
new branch. The first event is illustrated in Figure 9a. A
new circuit appears between the two sections shown in this
figure. Changes of this type have no effect on the value
of G_{max} or G_{min}. The appearance of a new branch is
illustrated in the serial sections shown in Figure 9b. Changes
of this type increase both G_{max} and G_{min} by one.

The connectivity of a subnetwork is <u>decreased</u> in two ways: by the disappearance of branches and by the collapse, or disappearance, of circuits. These events are the same as those illustrated in Figure 9 with the order of the serial sections reversed. The disappearance of a branch leaves both G_{max} and G_{min} unchanged. The collapse of a circuit decreases both G_{max} and G_{min} by one and leaves N unchanged unless the circuit which collapses was the last circuit associated with the appearance of a new circuit. In this case the presence of a separate part of the void space is indicated, as illustrated in the following paragraphs. The collapse of a circuit which was the last circuit associ-

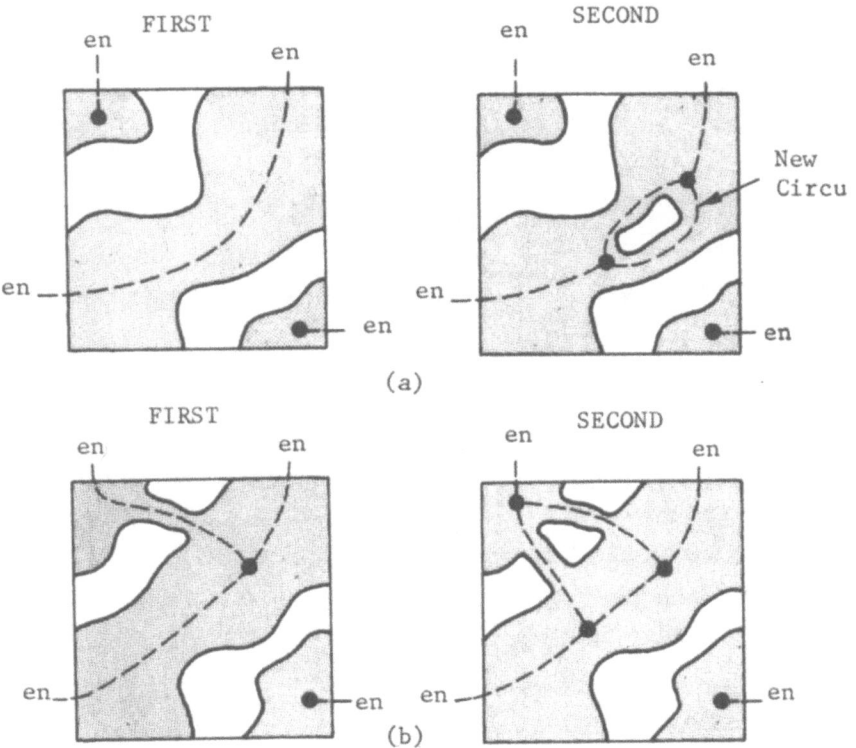

Fig. 9. Illustration of (a) appearance of a new circuit, and (b) appearance of a new branch, in the subnetworks constructed on successive serial sections.

ated with the appearance of a new circuit occurs only during
the analysis of the material space as separate parts of the
void space are observed while separate parts of the material
space are not observed. An event of this type leaves both
G_{max} and G_{min} unchanged and increases N by one.

During the analysis of the material space the void space
areas encircled by new circuits are denoted by a symbol,
such as X , as illustrated in Figure 10. When a circuit
collapses, due to the disappearance of a void area so

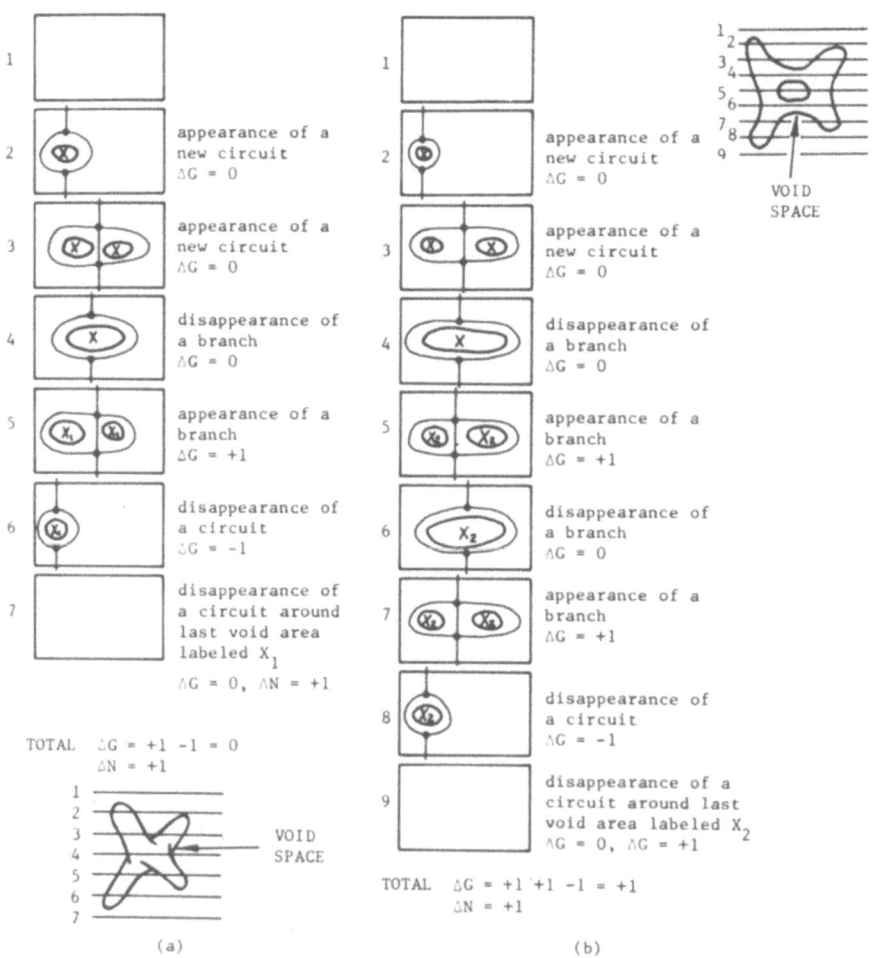

Fig 10. The appearance and disappearance of circuits in the
analysis of the solid space.

labeled, the presence of a separate part is indicated.

Figure 10a shows a simply connected separate part intersected by five serial sections which are also shown. The subnetworks are constructed on each serial section and the associated changes in the topological properties are indicated next to each serial section. The new void areas are labeled X . Two new void areas join to form one area between sections 3 and 4 and the resulting void area is labeled X . This void area then splits into two void areas between sections 4 and 5 both of which are labeled X_1 to indicate that they are joined on a previously analysed serial section. One of these void areas then ends between sections 5 and 6. The other void area labeled X_1 then ends between sections 6 and 7 and the circuit around this void area collapses. This event produces no change in G_{max} or G_{min} as the circuit which collapses between sections 6 and 7 was the last circuit associated with the appearance of a new circuit. An increase of one in the number of separate parts is associated with this event as indicated in Table 4.

Multiply connected separate parts can also be detected by this procedure as illustrated in Figure 10b. The labeling procedure is the same, X_2 is used here rather than X_1 to indicate that these new void areas are associated with different parts of the void space than those labeled X_1 . When the void area and circuit disappears between sections 8 and 9 the change in G_{max} and G_{min} is zero and the number of separate parts is increased by one because this is the last void area labeled X_2 . The total change in G_{max} and G_{min} is +1 as the separate part is multiply connected and has a genus of one.

The appearance of a new circuit and the appearance of a new branch are generally easily distinguishable as illustrated in Figures 9a and 9b. However, these events can appear to be very similar as shown by the following examples. Consider Figure 11b which shows two serial sections through the structure of Figure 11a. A branch is formed on the second serial section. The branch which appears is a circuit as it is incident on the same node at both ends.

However, the appearance of this branch is different from
the appearance of a circuit illustrated in Figure 9a. The
essential difference is that the circuit of Figure 9a can be
shrunk through the void space, in the third dimension, until
it collapses while the circuit of Figure 11b cannot be shrunk
until it collapses, even in the third dimension without crossing the pore-solid interface. Therefore, the circuit of Figure 9a is not an independent circuit and hence does not contribute to the genus while the circuit of Figure 11b is independent and hence does contribute 1 to the genus.

That the event illustrated in Figure 11b does represent
the appearance of a new branch can be easily seen when the
serial sections of Figure 11 are passed through the structure
at some other angle as illustrated in associated with different parts of the void space than those labeled X_1 . When

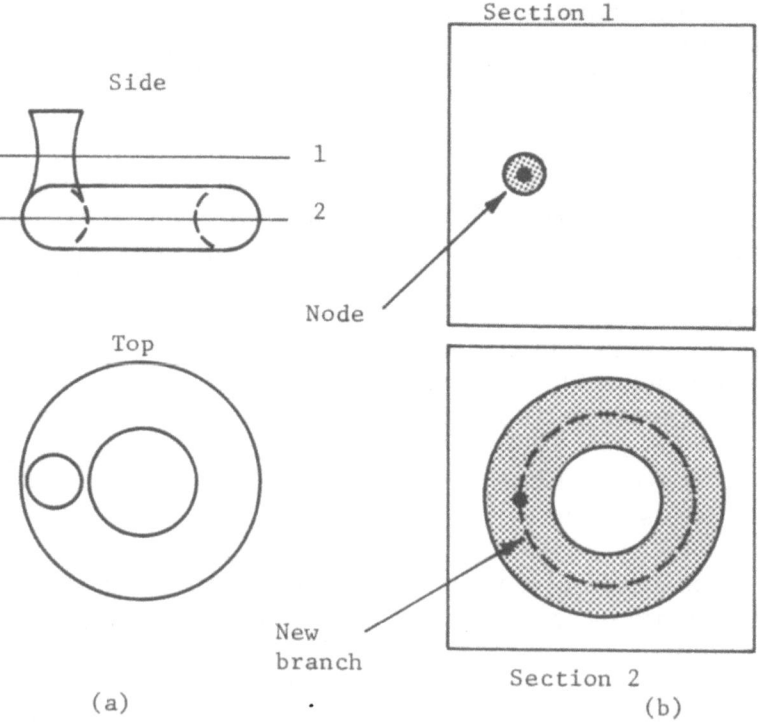

Fig. 11.   Illustration of the appearance of a new branch
contrasted with the appearance of a circuit.

the void area and circuit disappears between sections 8 and 9 the change in G_{max} and G_{min} is zero and the number of separate parts is increased by one because this is the last void area labeled X_2. The total change in G_{max} and G_{min} is +1 as the separate part is multiply connected and has a genus of one.

The appearance of a new circuit and the appearance of a new branch are generally easily distinguishable as illustrated in Figures 9a and 9b. However, these events can appear to be very similar as shown by the following examples. Consider Figure 11b which shows two serial sections through the structure of Figure 11a. A branch is formed on the second serial section. The branch which appears is a circuit as it is incident on the same node at both ends. However, the appearance of this branch is different from the appearance of a circuit illustrated in Figure 9a. The essential difference is that the circuit of Figure 9a can be shrunk through the void space, in the third dimension, until it collapses while the circuit of Figure 11b cannot be shrunk until it collapses, even in the third dimension without crossing the pore-solid interface. Therefore, the circuit of Figure 9a is not an independent circuit and hence does not contribute to the genus while the circuit of Figure 11b is independent and hence does contribute 1 to the genus.

That the event illustrated in Figure 11b does represent the appearance of á new branch can be easily seen when the serial sections of Figure 11 are passed through the structure at some other angle as illustrated in Figure 12a. It is apparent that a new branch is formed between sections 2 and 3.

Events such as that illustrated in Figure 11 occur infrequently during the analysis of actual sinter structures; however, events of this type should not be confused with the appearance of a new circuit.

The collapse of a circuit and the disappearance of a branch also must not be confused. If the order of the serial sections shown in Figures 12 and 9 are reversed then the

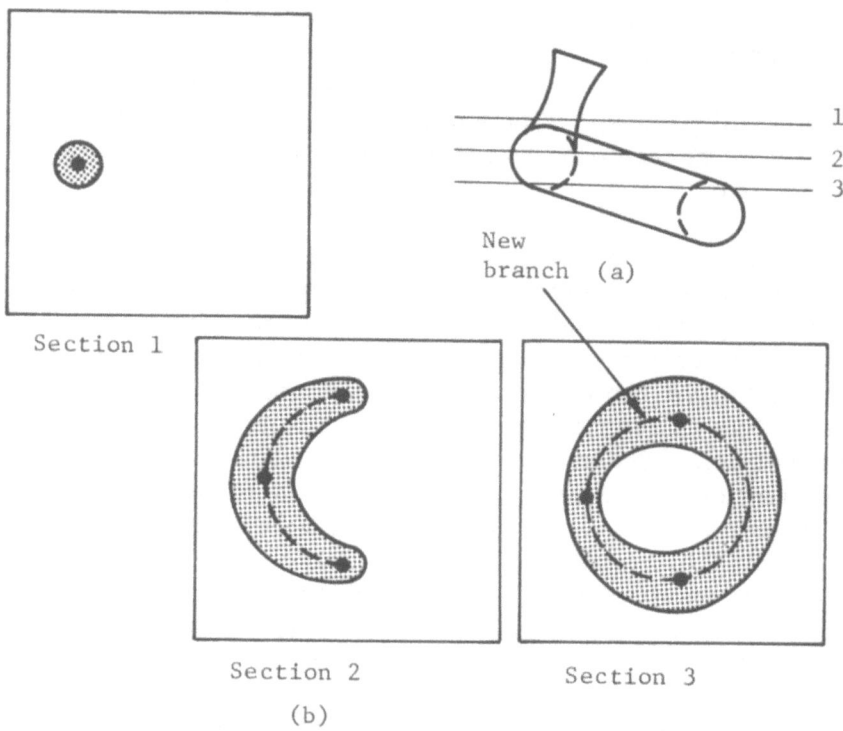

Fig. 12. The body in Figure 11 sectioned from a different orientation.

disappearance of a branch and the collapse of a circuit are illustrated, respectively.

Class III. Separations or New Connections between Subnetworks. Separations or connections of subnetworks between sections must be noted as successive sections are considered. This is facilitated by extending the labeling procedure. Subnetworks that result from the separation of a given subnetwork, are given the same label as the original subnetwork. For example, consider the serial sections shown in Figure 13. The subnetwork labeled 3 on the first section separates by the removal of a branch to form two subnetworks on the second section; both are labeled 3 according to the above convention. When two or more subnetworks with different labels join, the smallest of the labels is used to

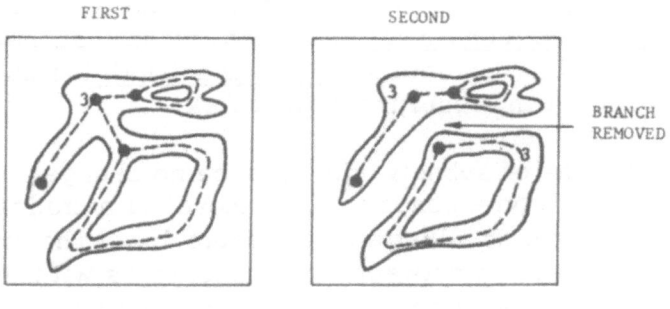

Fig. 13. Schematic illustration of the separation of a subnetwork between two successive serial sections.

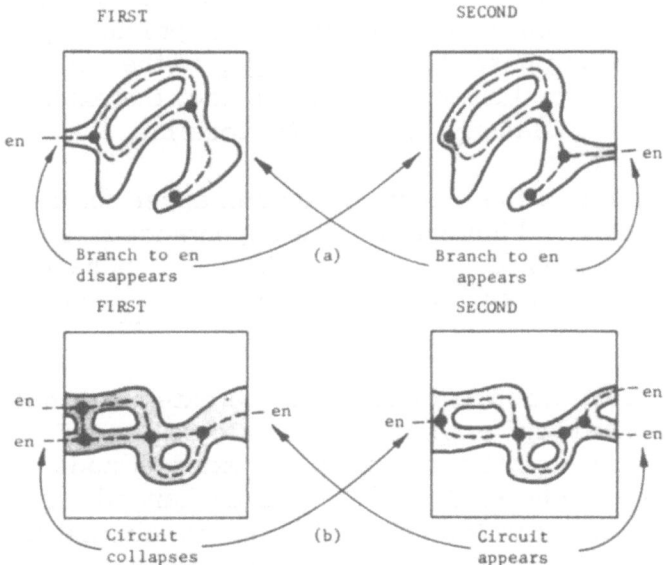

Fig. 14. Schematic illustration of sample surface effects in serial sectioning. The disappearance and appearance of (a) branches to the external node, and (b) circuits involving the external node are shown in successive serial sections.

relabel all subnetworks with the larger of the two labels on
the section under consideration. Also, when a subnetwork
labeled with a number joints a new subnetwork the number
is used to label the resulting subnetwork.

Separations of subnetworks produce no change in G_{max}
or G_{min}. This corresponds to the removal of a branch
between sections as illustrated in Figure 13a. The appearance of a connection between two subnetworks with the
same number produces an increase of one in both G_{max}
and G_{min}.

Joining events between numbered subnetworks and new
subnetworks and between two or more new subnetworks
leave G_{max} and G_{min} unchanged.

Class IV: Connections to the External Node. Changes
in the number of branches which contact the external node
must also be monitored. As pointed out previously only
branches which pass through the sides of the volume of
material analysed are assumed to contact the external node.
The following Class IV changes are possible.

Branches to the external node can appear or disappear
as illustrated in Figure 14a. The appearance of such a
branch increases G_{max} by one; G_{min} is unchanged. The
disappearance of such a branch leaves G_{max} and G_{min}
unchanged.

A circuit which is incident on the external node can
collapse or appear as illustrated in Figure 14b. The collapse
of a circuit which is incident on the external node produces
a decrease of one in G_{max}; G_{min} is unchanged. The appearance of a circuit which is incident on the external node
leaves G_{max} and G_{min} unchanged.

When the above tabulating procedure is applied to a
sequence of serial sections values of G_{max}, G_{min} and
N are obtained as each section is considered. The total
change in each of these parameters, after a given number of

sections has been analysed, is referred to as the "cumulative change" of that parameter. Only those events which produce changes in G_{max}, G_{min} and N are tabulated. Accordingly, events which must be noted are (see Table 4): disappearance of (a) subnetworks, (b) circuits, (c) branches to the external node, (d) circuits incident upon the external node; and appearance of (e) new branches, and (f) connections between subnetworks. The desired values of G_{Vmax}, G_{Vmin} and N_V are obtained by plotting the cumulative change in G_{max}, G_{min} and N vs. volume of material analysed and taking the slope of the linear portion of the respective curve, according to equation (22).

E. Rules of Thumb Concerning the Determination of the Topological Properties.

In order to obtain values for the topological properties with a minimum error and effort a number of factors must be taken into account. It will become apparent from consideration of these factors that the design of the experiment and the desired accuracy are closely related. It will also become apparent that it is not presently possible to obtain quantitative estimates of the various kinds of errors.

The accuracy of the result and the time spent on the analysis depends, to a large extent, on the size of the volume of material which is analysed. In general, as the volume analysed is increased the error is decreased and the time required to perform the analysis is increased.

1. The Effect of Sample Size on the Difference Between G_{Vmax} and G_{Vmin}. Given a microstructurally homogeneous material, the experimentally determined values of G_{Vmax} and G_{Vmin} bracket the actual value of genus per unit volume. The difference between these limits is equal to the number of branches incident on the external node divided by the volume of the sample analysed. The number of branches incident on the external node is directly proportional to the area of the sides of the sample. Thus the difference between G_{Vmax} and G_{Vmin} decreases as the ratio of the area of the sides to the volume of the sample analysed decreases.

Therefore, the area of each serial section chosen for analysis must be large enough so that the difference between G_{Vmax} and G_{Vmin} is sufficiently small for the purpose of the experiment.

It is found experimentally that if 150 to 200 features or void areas are contained in the area analysed on each serial section then the difference between G_{Vmax} and G_{Vmin} will be less than 15% of the value of G_{Vmax} [20]. In this case if G_{Vmax} and G_{Vmin} are averaged the value obtained will have less than about 7% error. This difference can be reduced further by analysing a larger area on each serial section.

2. Spacing of the Serial Sections. Some error is involved in comparing adjacent serial sections during the experimental analysis as it cannot always be unambiguously established that a certain event has occurred between sections. As a result it is necessary to make some arbitrary decisions during the analysis, the number of which increases, relative to the total number of observations, as the spacing between serial sections is increased. The error accompanying the correlation of adjacent serial sections may be reduced to an acceptable level by choosing the spacing of the sections so that the number of arbitrary decisions which must be made is a reasonably small fraction of the total value of the parameter being monitored. This spacing must be established for each structure examined. If a spacing of approximately 0.1 of the particle diameter is used then one event in about 250 has to be established arbitrarily.

As the spacing is decreased, a greater number of sections must be considered in order to analyse the same total thickness, or volume of material. Therefore, as the spacing is decreased, the total time spent in correlating the sections is increased proportionally.

3. Total Number of Serial Sections. The number of sections which must be analysed before the linear portion of the cumulative plots is reached depends on the circuit size or the size of the separate parts in the structure. Linearity is not reached until a thickness has been analysed such that

the largest circuits or separate parts in the sample are detected. After linearity is reached an additional number of
sections must be analysed in order to determine the slope
of the linear portion. This number of sections can be
established by applying a least squares analysis to the
linear portion of the curve. It has been found experimentally
that less than 50 serial sections are necessary in order to
establish the slope to within a few per cent error.

4. Structural Inhomogeniety. The structure of sintered
materials is inhomogeneous, at least over short distances,
as a result of irregularities in the particle stacking, and
over greater distances in compacted structures. If a sample
containing several hundred particles is analysed, small
scale inhomogenieties are averaged out. Any long range
variation in the topological properties can, in principal, be
detected by analysing samples at various locations in the
sinter body. However, at the present time the effort required
to determine the topological properties for a single sample
of the structure would make the quantitative characterization
of long-range inhomogenieties in a single compact a time
consuming undertaking.

5. Error Resulting from the Uncertainty in the Position
of the Deformation Retract. The determination of the genus
and number of separate parts is based, in principle, on
consideration of the deformation retract or network represen-
ting the void or material space of the sinter structure. The
position of the retract can not be uniquely established and
its position become increasingly uncertain as the volume
fraction increases. As a result the error introduced by this
effect increases as the volume fraction of the phase analysed
increases. For this reason the phase with the smallest
volume fraction, void or solid, should be chosen for analysis.

A quantitative estimate of this error is difficult to obtain. At present it can only be stated that if one to two
hundred branches are contained in the area analysed on each
serial section, the error produced by this effect is less
than a few per cent.

6. <u>Error Due to Large Scale Features in the Structure</u>.
Large circuits or separate parts in the structure under consideration may not be observed due to the limited size of
the volume of structure which is analysed. The smallest
loops or separate parts which can remain unobserved are
approximately the size of the volume analysed. This effect
becomes more important for small values of genus per unit
volume. In this case larger volumes must be analysed in
order to establish a result of the same accuracy.

Once a sample has been analysed, in which the thickness is of the order of the dimensions of the area analysed
on each section, further increase in the thickness analysed
will not increase the accuracy of the results. Larger loops
or separate parts will not be detected simply by increasing
the thickness analysed. In order to further increase the
accuracy, both the area and thickness analysed, must be
increased.

The error produced by this effect is negligible when
the area analysed on each serial section contains one to
two hundred branches or features and a thickness equivalent to 4 to 5 particle diameters is analysed.

7. <u>Separate Parts Error</u>. A maximum and minimum limit
is not obtained for the number of separate parts per unit
volume, as is done for the genus per unit volume. As a
result separate parts which intersect the surface of the
volume analysed are not taken into account by the method
of analysis presented in Section D.

It is observed experimentally that isolated separate
parts of the void space in sinter structures are predominantly small, about 90% being less than one half the diameter
of the particles from which the structure was produced[20].
As a result, the ratio of the number of separate parts which
intersect the surface of the volume analysed, to those observed inside the volume analysed, is small. Therefore,
when separate parts which intersect the sides of the volume
analysed are neglected the resulting error is small and

decreases as the surface to volume ratio of the sample analysed decreases. Therefore, as in the determination of the genus, it is desirable to analyse as large an area as practical on each serial section.

The separate parts which intersect the sides of the volume analysed can in principle be taken into account as indicated in Section D. This is done by observing a volume of material somewhat larger than the sample analysed. Portions of the void space which intersect the sides of the sample may then be separately analysed in order to determine whether or not they are separate parts.

The latter procedure should be followed when it is necessary to count all separate parts, or when there is special interest in the larger separate parts in the structure.

8. Summary of the Design of Serial Sectioning Experiments. As a result of the various errors inherent in the determination of the topological properties the following summary can be made concerning the experimental procedure and the accompanying accuracy of the result. In general the space with the lowest volume fraction should be chosen for analysis. The serial sections should be spaced so that about 10 intersect each particle. The area analysed on each serial section should contain about 150 features or void areas and about 50 serial sections should be analysed. Under these conditions the average value of $G_{V_{max}}$ and $G_{V_{min}}$ will be within about 10% of the actual value of genus of the structure analysed.

IV. APPLICATIONS TO SINTERED STRUCTURES

The measurements outlined in the previous two sections were applied to a wide range of sintered structures, exploring the effects of particle size, stacking, shape, precompaction, sintering temperature, and material. These results are reported elsewhere in detail[10]. In order to show the results of the quantitative approach to studying sintering, and to demonstrate the insight this approach gives to the

analysis of the process, an example of the complete characterization of the evolution of microstructure is presented in
this section.

A spherical, oxygen free atomized copper powder was
chosen for this example. A single size fraction, -270 +
325 mesh or 48 microns, was sintered to a range of densities
in dry hydrogen at 1000°C. The resulting series of samples
covered the full range of porosity levels available to this
powder. These samples were examined in reflection microscopy and by serial sectioning. All of the topological and
metric properties were determined as a function of the
volume fraction of porosity in the system.

A. The Metric Properties.

The metric properties determined from the counting
measurements outlined in Section II included: (1) the pore
volume fraction, V_V^p; (2) the surface area of the pore-solid
interface, S_V^{ps}; (3) the surface area of grain boundaries,
S_V^{ss}; (4) the total curvature of the pore-solid interface, M_V^{ps};
(5) the mean pore intercept, $\overline{\lambda}$; (6) the mean solid intercept,
$\overline{\lambda}_s$; (7) the mean grain intercept, $\overline{\lambda}_g$; and (8) the average
mean surface curvature $\overline{H}$, of the pore-solid interface. Plots
of several of these properties versus the level of porosity
in the structure are presented in Figure 15.

B. The Topological Properties.

The serial sectioning technique described in Section III
was applied to a set of samples selected because they were
uniformly distributed along the V_V^p axis. Both the genus
and number of separate parts of the pore-solid interface were
quantitatively determined. The results of these studies are
shown in Figure 15.

C. The Evolution of Microstructure.

Examination of the data presented in Figures 15 has
revealed that the sintering process is divided into three

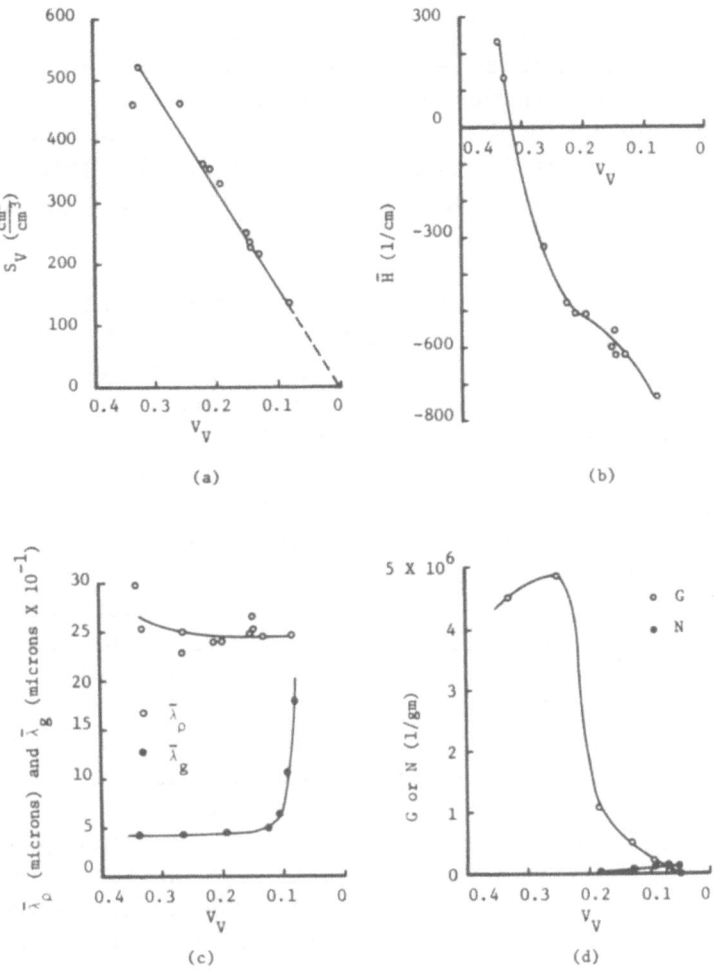

Fig. 15. Paths of microstructural
change for a −270 +325 mesh (48 micron)
size fraction of spherical copper
powder, sintered in dry hydrogen at
1000°C.

stages, each with its own characteristic unit geometric process and distinctive path of structural change.

The first stage of sintering is characterized by the smoothing of short range surface irregularities, and the growth of interparticle welds or contacts. The genus remains nearly constant, while the surface area decreases, and the average curvature decreases toward zero. The grain size remains essentially constant. The dominant unit geometric change is neck growth.

In the second stage, the genus decreases rapidly toward zero. A small number of portions of the porosity become isolated into separate parts. The mean curvature becomes negative, and continues to increase negatively. The surface area decreases linearly with the volume fraction of porosity, indicating that a balance has been achieved between transport mechanisms which favor surface rounding, and those which promote densification[10]. The mean pore intercept remains essentially constant, as does the mean grain intercept. The dominant unit geometric process is the constriction and closure of channels within the pore network; this process is responsible for the topological changes that occur.

The third stage of sintering is characterized by the isolation of porosity as the channel closure process achieves completion. The decrease in surface area is no longer linear, and the mean curvature begins to decrease (become less negative). The mean pore intercept increases somewhat, and the mean grain intercept suddenly increases sharply. The dominant geometric process in third stage sintering is coarsening of isolated pores.

Thus, the quantitative study of sintering clearly reveals that the process is divided into three distinct stages, each dominated by a different unit geometric change in which the sources and sinks for the mass transport which produces those changes are distinctly different. Accordingly, the kinetics of the process cannot be described in terms of a single kinetic equation, or a single mechanism.

V. SUMMARY

A complete, quantitative characterization of the microstructure of a porous body is made available by the application of the techniques of quantitative microscopy. This
characterization may be formulated in terms of properties
that have precise geometric meaning, even for so complex a
structure as a multiply connected pore network. Properties
which may be measured include:

1. The metric properties;
    a. pore volume fraction,
    b. surface area of the pore-solid interface,
    c. surface area of grain boundaries,
    d. total curvature of the pore-solid interface,
    e. total length of grain edge,
    f. total length of ssp triple lines,
    g. mean intercept lengths of porosity, solid,
        and grains,
    h. average mean surface curvature;
2. The topological properties;
    a. genus of the pore-solid interface,
    b. connectivity of the pore network,
    c. number of separate parts of the pore-solid
        interface,
    d. number of separate, isolated pores.

Application of these measurements, in whole or in
part, to real sintered structures provides quantitative information about the evolution of microstructure during sintering,
the effect of powder and processing variables upon the structural evolution, and relationships between microstructure
and mechanical and physical properties. This information
may, in turn, provide a quantitative basis for the assess-
ment of theoretical descriptions of mechanisms of sintering,
and of the properties of sintered structures, and ultimately
may lead to the design of sintering processes to produce
sintered structures with preselected combinations of
properties.

REFERENCES

1. R. T. DeHoff and F. N. Rhines, Editors, Quantitative Microscopy, McGraw-Hill, New York, 1968.

2. J. W. Gibbs, The Scientific Papers of J. Willard Gibbs, Dover Publications, New York, 1961, p. 229.

3. R. L. Fullman, Trans. AIME, 197, 1953, p. 447.

4. F. Chayes, Petrographic Modal Analysis, John Wiley & Sons, Inc., New York, 1956.

5. J. E. Hilliard, in Quantitative Microscopy, R. T. DeHoff and F. N. Rhines, Editors, McGraw-Hill, New York, 1968, Chapter 3.

6. C. F. Fisher and M. Cole, The Microscope, 16, 1968.

7. H. F. Fischmeister, in Quantitative Microscopy, R. T. DeHoff and F. N. Rhines, Editors, McGraw-Hill, New York, 1968, Chapter 12.

8. C. S. Smith and L. Guttman, Trans. AIME, 197, 1953, p. 81.

9. R. T. DeHoff, Trans. Met. Soc. AIME, 239, 1967, p. 617.

10. F. N. Rhines and R. T. DeHoff, Second European Symposium on Powder Metallurgy, Max Planck Inst., Stuttgart, Germany, 1968, p. 1.

11. P. G. Hoel, Introduction to Mathematical Statistics, John Wiley & Sons, Inc., New York, 1954, p. 107.

12. J. E. Hilliard and J. W. Cahn, Trans. Met. Soc. AIME, 221, 1961, p. 344.

13. K. Craig, Masters Thesis, University of Florida, 1966.

14. J. H. Steele, Jr., Doctoral Dissertation, University of Florida, 1967.

15. R. W. Kraft, F. D. Lemkey, and F. D. George, Trans. Met. Soc. AIME, 224, 1962, p. 1037.

16. J. Kronsbein, L. J. Buteau, Jr., and R. T. DeHoff, Trans. Met. Soc. AIME, 233, 1965, p. 1961.

17. R. T. DeHoff, in Quantitative Microscopy, R. T. DeHoff and F. N. Rhines, Editors, McGraw-Hill, New York, 1968, Chapter 10.

18. L. K. Barrett and C. S. Yust, ORNL Report, No. 441, 1969.

19. M. L. Pickelsimer, in Quantitative Microscopy, R. T. DeHoff and F. N. Rhines, Editors, McGraw-Hill, New York, 1968, Chapter 11.

20. E. H. Aigeltinger, Doctoral Dissertation, University of Florida, 1970.

TECHNIQUES FOR THE STUDY OF HOMOGENIZATION IN COMPACTS OF
BLENDED POWDERS

R. W. Heckel
R. D. Lanam
R. A. Tanzilli

Department of Metallurgical Engineering
Drexel University
Philadelphia, Pa.   19104

I. INTRODUCTION

The production of alloys by powder metallurgy may be
accomplished by conventional powder processing of prealloyed
powders or by the processing of compacted blends of powders.
The latter method necessarily requires the insertion of a
homogenization step in the processing sequence. In general,
the homogenization is achieved through solid-state interdiffusion of the components in the compacted blend.

In spite of the fact that the processing of compacted
blends of powders requires an extra processing step (homogenization), there are certain advantages to this production
method:
   a. a wide variety of alloys may be produced with a
small number of different powders by varying the ratio of
the blended powders,
   b. alloys with unique microstructures and properties
may be fabricated by suitable control of the degree of homogeneity,
   c. high densities and green strengths may be achieved
during compaction of blended powders if the major component
is present as a soft, ductile elemental powder,
   d. deformation subsequent to compaction and sintering
can possibly be carried out more easily by control of the
degree of homogeneity.

These advantages, however, must be weighed against the possible disadvantages of processing with a homogenization step:
    a.   a high-temperature, long-time homogenization treat-
ment may be necessary if the particle sizes are large,
    b.   small particle sizes necessary to provide rapid
homogenization may introduce objectionable amounts of oxide
into the compact from particle surface contamination,
    c.   the possible Kirkendall porosity resulting from the
homogenization treatment may be objectionable,
    d.   the formation of brittle, intermediate phases during
homogenization may limit subsequent deformation processing.
The choice of homogenization processing for the production of
a particular alloy must be arrived at after consideration of
the above-mentioned advantages and disadvantages.

Numerous studies on the production of alloys from powder
blends have been described in the literature. In many of
these studies, little attention has been given to either the
theoretical prediction or experimental determination of the
homogeneity achieved by the processing steps that were used.
Often, the physical and mechanical properties of the powder-
fabricated alloy are simply compared to handbook or experimental values of cast and wrought alloys. Large variations
in such comparisons may not be the result of the powder method
itself, but may result from inadequate homogeneity.

Over the past ten years, significant advances have been
made in the understanding of homogenization behavior of powder
compacts. These advances have resulted from newly-developed
experimental methods for observing compositional heterogeneities in alloys. In addition, newer high-speed computer facil-
ities have led to the use of numerical methods for solving the
complex diffusion problems characteristic of powder homogenization. It is the purpose of this paper to review the applicability of these developments to the problem of homogenization
of powder compacts. In particular, this paper is aimed at:
    a.   presenting mathematical models that are applicable
to a wide variety of powder homogenization problems,
    b.   presenting direct methods for quantitative deter-
mination of the degree of homogeneity in powder compacts,
    c.   showing comparisons of results obtained from model
predictions and data obtained from quantitative techniques.

II. THE HOMOGENIZATION PROCESS

A. Compacted Blends of Powders

In general, compacted blends of powders may be characterized by particles of one composition dispersed in a continuous matrix of a different composition. The amount of
interparticle porosity in the compact is a function of the
compaction pressure and the mechanical properties of the
powders. The matrix is composed of the softer and/or more
abundant particles in the blend. In many instances, the
geometry of the compacted blend may be idealized to spherical
particles of the minor constituent embedded in a continuous

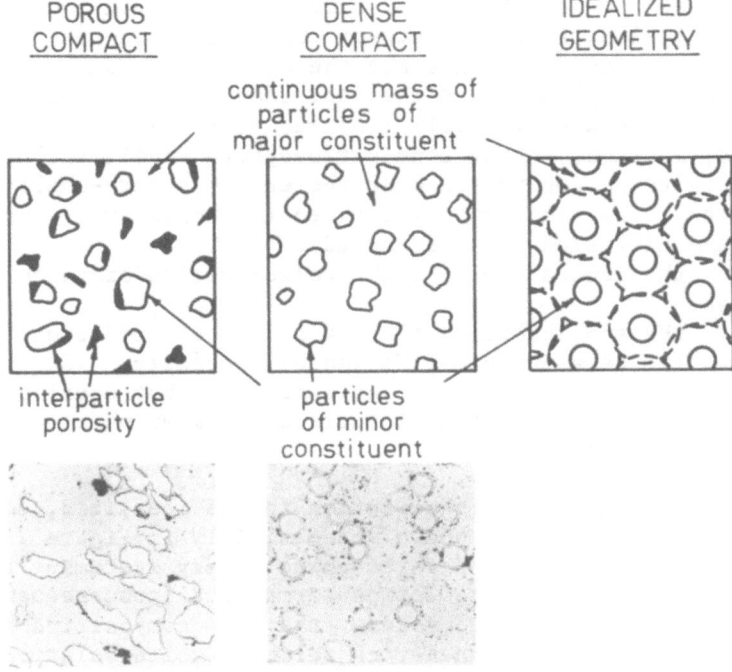

Figure 1. Schematic representation of porous and dense compacts along with the idealized geometry of the concentric-
sphere model. The molybdenum particles in a nickel matrix
(lower left) constitute a porous compact; the tungsten particles in a nickel matrix (lower center), a high density
compact. (Ni-Mo and Ni-W compacts were given a low-temperature
sinter to facilitate metallographic preparation.)

matrix of the major constituent. This situation is shown in
Figure 1. The idealized geometry has been referred to as the
concentric-sphere model wherein each minor constituent
particle has associated with it a shell of matrix. The
mean composition of each sphere-shell composite is the same
as the entire compact, facilitating mathematical modeling.
The homogenization behavior of the concentric-sphere model
is, therefore, independent of the particle size of the major
constituent.

B. Homogenization of One-Phase, Two-Phase, and Three-Phase Systems

The homogenization behavior of powder compacts that
approximate the idealized, concentric-sphere geometry may
be compared to the behavior of ordinary diffusion couples.
The principal differences between the concentric-sphere
model and diffusion couples are the diffusion geometry
(spherical vs. planar) and the fact that diffusion distances
in couples are usually small compared to the size of the
couples, whereas, in the concentric-sphere model, the diffusion is to be considered over the entire sphere-shell
composite. The similarities between the concentric-sphere
model and diffusion couples include:
a. the general form of the concentration-distance
profiles,
b. the applicability of Fick's laws of diffusion,
c. the development of Kirkendall porosity,
d. the validity of the approximation that equilibrium
compositions are present at interfaces between phases in
multiphase systems.

The progress of homogenization in the idealized, concentric-sphere model is shown schematically in Figure 2 for
one-phase, two-phase, and three-phase binary systems. (This
figure will also serve to define terms used in subsequent
sections.) In all three systems, the mean composition of
the blend, $\bar{C}$, lies in the A-rich terminal solid solution.
For situations where the differences in density:molecular
weight ratios of A and B are small, $\bar{C}$ (atom fraction)* may
be approximated by:

$$\bar{C} = \left(\frac{\ell}{L}\right)^3 \tag{1}$$

---

* Concentrations are atom fractions unless otherwise noted.

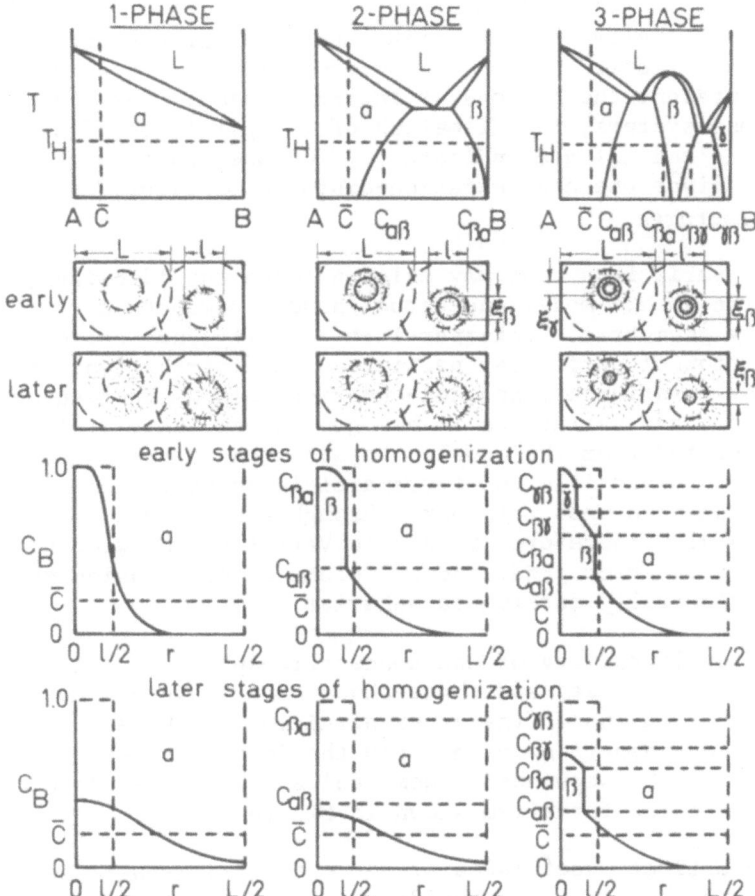

Figure 2. Schematic representation of the homogenization process in one-, two-, and three-phase binary systems in terms of microstructural changes and composition-distance profiles.

where ℓ is the diameter of the minor constituent particle and L is the diameter of the composite. In addition, if the number of minor constituent particles per unit volume, N_V, is known, L can be approximated from:

$$N_V \stackrel{\sim}{=} \frac{1}{\frac{4}{3}\pi\left(\frac{L}{2}\right)^3} \qquad (2)$$

Schematic representations of microstructures are shown in Figure 2 (below the appropriate phase diagrams) for both "early" and "later" stages of homogenization. The shaded areas represent the regions of interdiffusion. For the multiphase systems, the diameters of the particles of the phases (ξ_β and ξ_γ) are indicated. The lower portion of Figure 2 gives the concentration-distance profiles for the microstructures.

The variations that may exist between the idealized, concentric-sphere model and actual powder compacts are primarily geometric. Interparticle porosity, non-uniform distribution of particles, particle size distributions of the minor constituent, and non-spherical minor constituent shape can cause variations in homogenization behavior from that predicted from the model. In addition to these geometry effects, interparticle and intraparticle porosity can give rise to enhanced mass transport through the compact. These effects would, therefore, result in variations between homo-genization behavior predicted by volume interdiffusion models and experimentally-determined behavior.

The applicability of the concentric-sphere geometry to one-phase, two-phase, and three-phase binary systems is shown in Figure 3. Although the minor particle (Ni, W, and Mo) distributions depart somewhat from the idealized geometry of Figure 1, the sequence of homogenization events proceeds according to the formalism shown in Figure 2.

The ultimate usefulness of the concentric-sphere geometry lies in the fact that it provides a reasonable framework for diffusion models that can describe homogenization. Such models, although idealized, have value in that they provide:
   a. a basis for systematic consideration of homogeniza-tion phenomena,
   b. a prediction of homogeneity after a specific heat treatment,
   c. predictions of the changes that would occur in homo-geneity when processing conditions are altered,
   d. a basis for comparison of experimental data.
In general, models cannot account for the effects of porosity, size distributions, poor mixing, and irregular particle shape as discussed previously. However, if the models are used as standards for comparison, the effects of non-idealities may be evaluated. Thus, models provide a reference point which may be considered to be as useful an experimental method as quantitative heterogeneity measurement techniques.

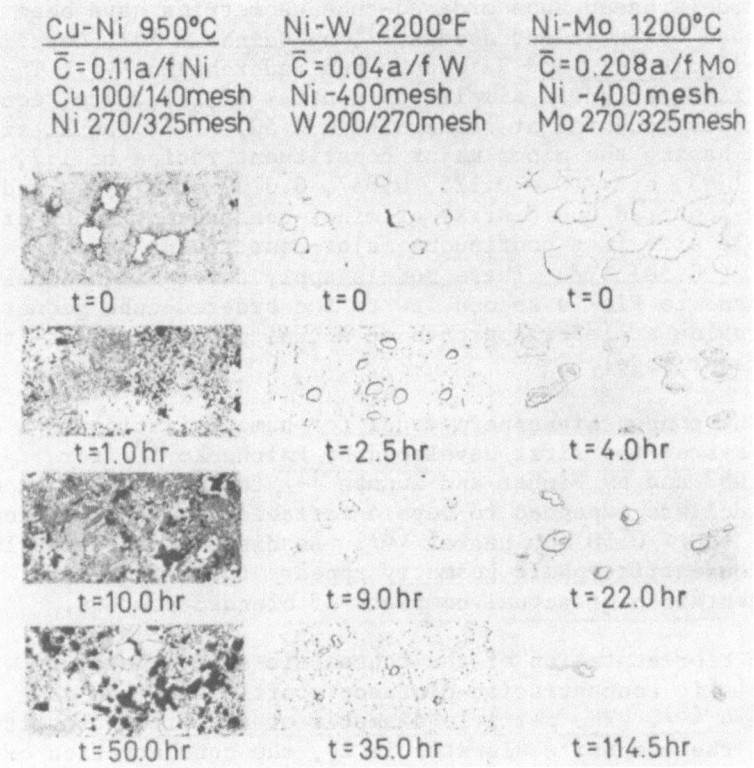

Figure 3.   Photomicrographs of the homogenization process in
one-phase (Cu-Ni), two-phase (Ni-W), and three-phase (Ni-Mo)
systems.   The Ni-Mo compacts were homogenized under an axial
stress of 1000 psi. to minimize the formation of Kirkendall
porosity.

III.   HOMOGENIZATION MODELS

A. One-Phase Binary Systems

Models of the one-phase, binary, homogenization process
shown in Figure 2 have been introduced by a number of authors.
Chevenard and Waché [2] and Duwez and Jordan [3] have described a model for the homogenization of alternate plates
(planar model).   These authors solved Fick's second law for
the situation where all of the plates had the same thickness,
making the model applicable to alloys near $\bar{C} = 0.50$.

Models based upon ordered-cube geometries have been
developed by Duwez and Jordan[3], by Weinbaum [4], Gertzriken
and Feingold [5], and Raichenko and Fedorchenko [6]. These
geometries include a simple cubic array of alternate, equal-
size cubes touching at corners ($\bar{C}$ = 0.50) [3,4], equal-size
arrays having the minor:major constituent ratios of 1:7,
1:26, 1:63, etc. ($\bar{C}$ = 0.125, 0.037, 0.016, etc.) [5], and
a face-centered cubic array of minor-constituent cubes of
variable size in a continuous major-constituent matrix
(0 < $\bar{C}$ ≤ 0.50) [6]. These models apply three-dimensional
solutions to Fick's second law to the ordered-cube geometries
and provide a closer approach to actual powder compacts than
the planar model.

The concentric-sphere model for homogenization in a one-
phase system was first developed by Raichenko [7] for
$\bar{C}$ = 0.037 and by Fisher and Rudman [1] for $\bar{C}$ = 0.50. Later,
the model was expanded to have a variable mean composition
(0.001 ≤ $\bar{C}$ ≤ 0.50) by Heckel [8]. As discussed previously,
this concentric-sphere geometry appears to be the best
representation of actual compacts of blended powders.

A representation of the concentric-sphere geometry with
a schematic concentration-distance profile is given in
Figure 4 [8]. The particle diameter of the minor constituent
is d_A, the composite diameter is 2b, the concentration of the
minor constituent is C_A, and $\tau = Dt/d_A^2$, where D is the con-
centration-independent interdiffusion coefficient and t is
the homogenization time. The value of C_A was found as a
function of r/b and t by numerical solution of Fick's second
law in spherical coordinates:

$$\frac{\partial C_A}{\partial t} = D \left(\frac{\partial^2 C_A}{\partial (r/b)^2} + \frac{2}{(r/b)} \cdot \frac{\partial C_A}{\partial (r/b)} \right) \quad (3)$$

with the initial conditions:

$$C_A = 1.0 \qquad 0 \le r < a$$

$$C_A = 0 \qquad a < r \le b$$

and boundary conditions:

$$\frac{\partial C_A}{\partial r} = 0 \qquad r = 0; \quad r = b$$

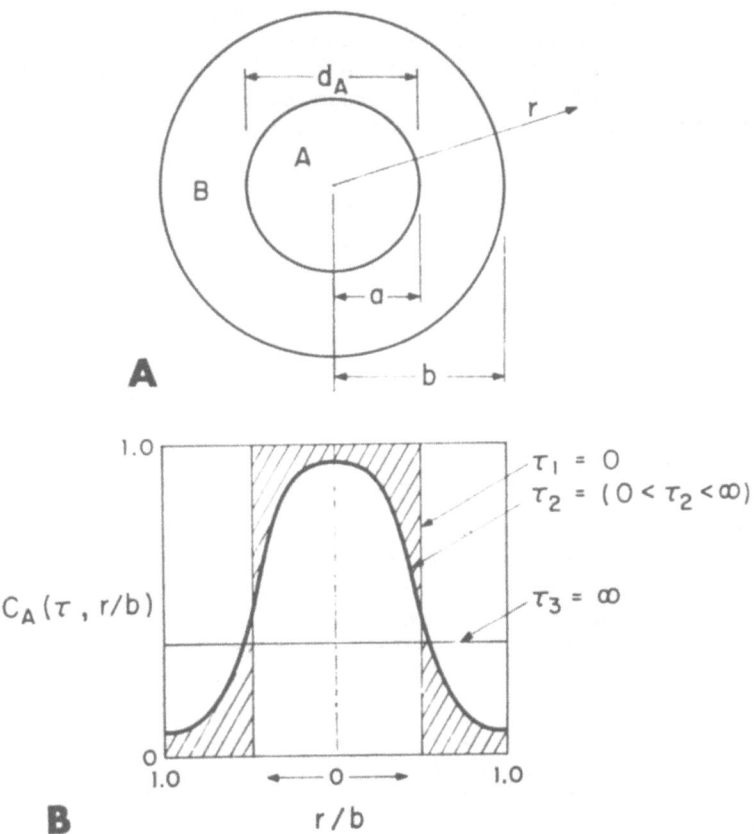

Figure 4. Schematic representation of (A), the concentric-sphere geometry, and (B), the concentration profiles through the concentric-sphere model for various values of time (τ). Since spherical symmetry elements of constant size cannot be stacked to fill space, the meaning of the dimension, b, may be considered in terms of the approximation given by Equation 2. The crosshatched areas on either side of the surface r = a represent the mass of material transferred across this boundary at time, τ_2. (Heckel [8]; courtesy of the Transactions of ASM).

The mean composition was varied by:

$$\bar{C} = \left(\frac{a}{b}\right)^3 \tag{4}$$

which is equivalent to Equation 1. Following Fisher and
Rudman [1], the approach to complete homogeneity was monitored by the degree of interdiffusion, F, which was determined by the ratio of the mass crossing the surface r = a
up to a given value of t to that crossing the surface in
infinite time:

$$F = \frac{m_t}{m_\infty} \tag{5}$$

where:

$$m_t = 4\pi \int_0^{a/b} (1-C_A) \left(\frac{r}{b}\right)^2 d\left(\frac{r}{b}\right) + 4\pi \int_{a/b}^1 C_A \left(\frac{r}{b}\right)^2 d\left(\frac{r}{b}\right) \tag{6}$$

and

$$m_\infty = \frac{8\pi}{3} \bar{C} (1-\bar{C}) \tag{7}$$

The degree of interdiffusion, F, is, therefore, a normalized
homogenization parameter which increases from zero (t = 0)
to unity (for complete homogenization). The range of compositions existing in the composite shown in Figure 4 was also
monitored as a means of following homogenization. The minimum,
$C_{A\ min.}$, occurs at r/b = 1.0 and the maximum, $C_{A\ max.}$, occurs
at r/b = 0. The variation of F, $C_{A\ min.}$, and $C_{A\ max.}$ as a
function of τ and $\bar{C}$ are shown in Figure 5 [8]. The results
presented in this figure represent the concentric-sphere model
for one-phase homogenization.

Figure 5 may also be applied to the more general initial
condition where a compacted blend of two different prealloyed
powders are homogenized:

$$C_A = C_2 \neq 1.0 \qquad 0 \leq r < a$$

$$C_A = C_1 \neq 0 \qquad a < r \leq b$$

where both C_1 and C_2 lie in the same phase field. To apply
Figure 5 to this situation, compositions C_1, $\bar{C}$, and C_2 must
be normalized on a concentration scale from zero to unity.

Often, the use of Figure 5 can be awkward due to preliminary calculations necessary to obtain τ. The nomograph in

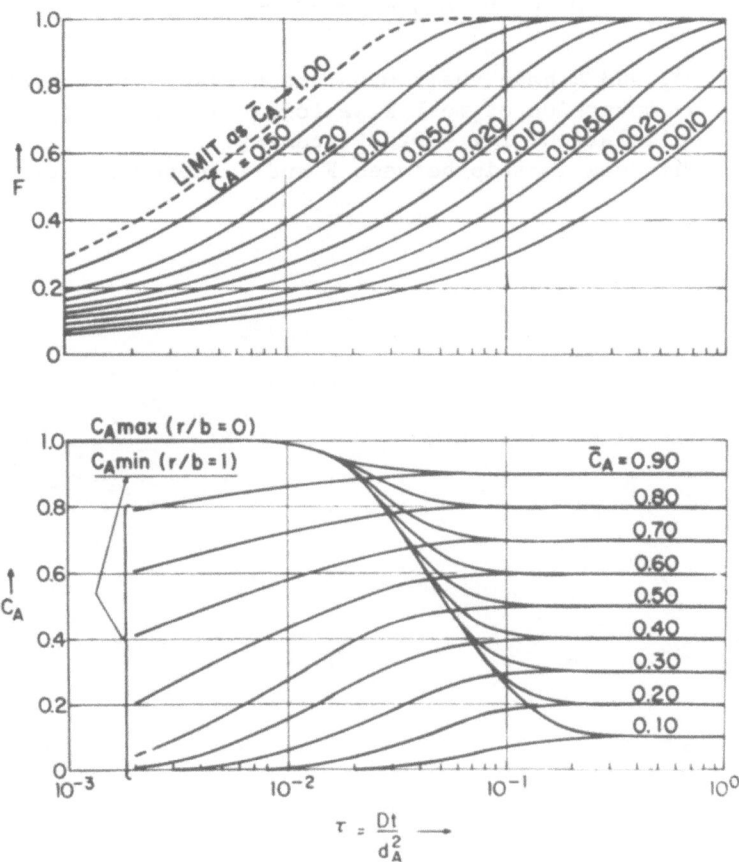

Figure 5. Variation of the degree of interdiffusion, F, and the range of composition, $C_{A\ max.}$ and $C_{A\ min.}$, as functions of mean composition, $\bar{C}$, and τ, according to the concentric-sphere model. (Heckel [8]; courtesy of the Transactions of ASM).

Figure 6[(8)] is designed to provide:
    a. a graphical calculation of D for a particular homo-
genization temperature, T ($^\circ$C), from:

$$D = D_o \exp\left(- \frac{Q}{R(T + 273)}\right) \tag{8}$$

where D_0, Q, and R have their usual meanings (upper left),
    b. a conversion of mesh size to d_A (lower right),
    c. calculation of $\tau = Dt/d_A^2$ (center),
    d. the relationship between F and τ (upper right).

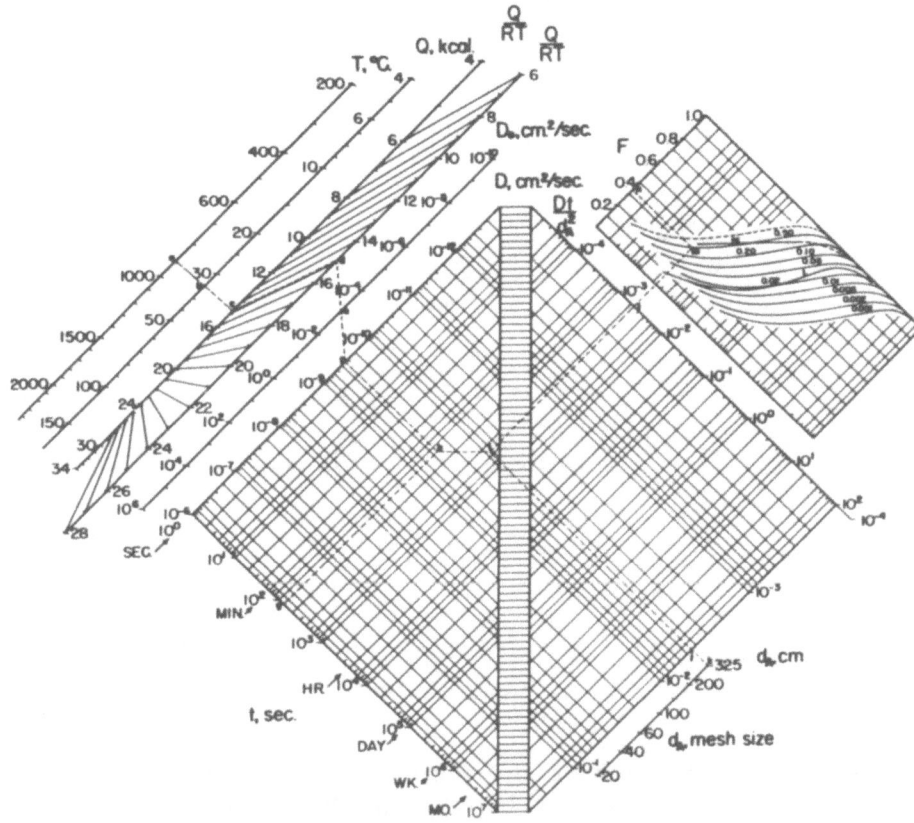

Figure 6. Nomograph for providing a graphical calculation
of the degree of interdiffusion, F, for the specific processing parameters D, d_A, $\bar{C}$, and t. The upper left portion of
the figure provides a graphical calculation of D in terms of
D_0, Q, and T ($^\circ$C). (Heckel [(8)]; courtesy of the Transactions
of ASM).

Figure 6, therefore, provides a complete graphical representation of the concentric-sphere, one-phase, homogenization model in terms of the basic input processing parameters.

Comparison of the planar (laminar), ordered cube, and concentric-sphere models for $\bar{C}$ = 0.50 shows a wide variation in predicted homogenization behavior. Figure 7 indicates

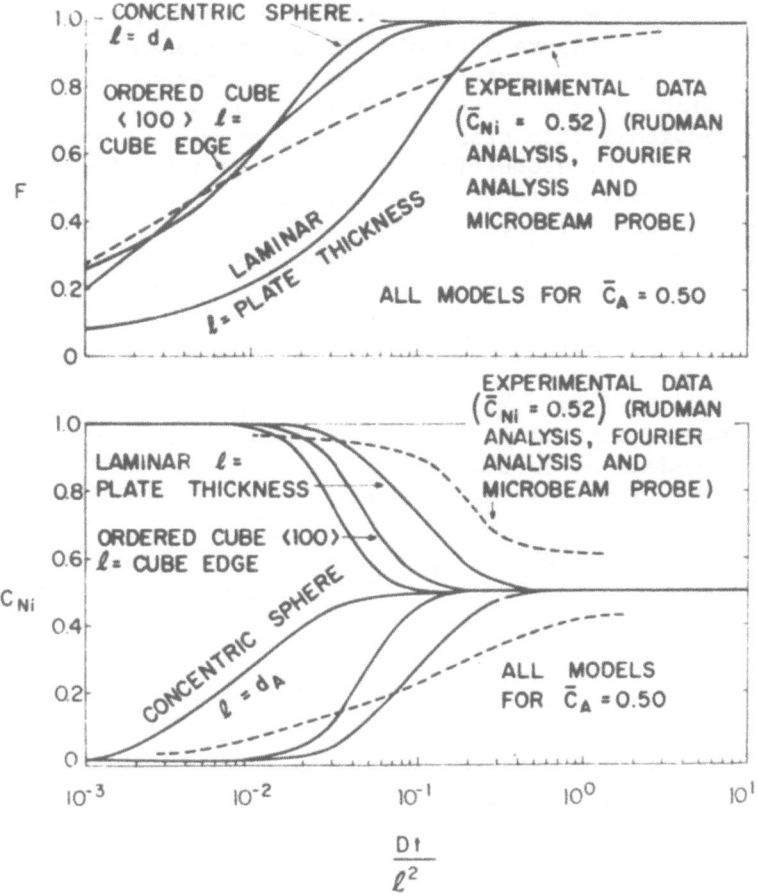

Figure 7. Comparison of the concentric-sphere model with the ordered-cube and laminar homogenization models. The degree of homogenization, F, and the range of concentrations, $C_{A \text{ max.}}$ and $C_{A \text{ min.}}$, calculated from all three models for $\bar{C}$ = 0.50 are compared. Also shown are experimental data for mixed elemental copper and nickel powders having approximately the same mean composition ($\bar{C}$ = 0.52). (Heckel [8]; courtesy of the Transactions of ASM).

that the planar (laminar) model predicts much slower homogenization than the other models. Although the ordered-cube
and concentric-sphere models compare favorably in degree of
interdiffusion, F, the ordered cube model predicts slower
homogenization on the basis of the range of compositions.

B. Two-Phase Binary Systems

The two-phase, binary homogenization process shown
schematically in Figure 2 results from the presence of the
moving heterophase interface and the necessity of treating
the diffusion process in each of the two phases. The solution
to this homogenization problem has been given by Tanzilli and
Heckel [9] for the geometries given in Figure 8. The numerical
methods and computer techniques employed in their treatment
involved the simultaneous solution of:

    a.　Fick's second law in both the α and β phases:

$$\frac{\partial C}{\partial t} = \frac{1}{x^m} \frac{\partial}{\partial x} \left(x^m D \frac{\partial C}{\partial x} \right) \tag{9}$$

where m = 0, 1, 2 for planar, cylindrical, and spherical
geometries, respectively, x is the distance parameter, and
D is independent of concentration,

    b.　the interface mass balance:

$$(C_{\beta\alpha} - C_{\alpha\beta}) \frac{d(\xi/2)}{dt} = D^\alpha \left(\frac{\partial C^\alpha}{\partial x} \right)_{x = \frac{\xi^+}{2}}$$
$$- D^\beta \left(\frac{\partial C^\beta}{\partial x} \right)_{x = \frac{\xi^-}{2}} \tag{10}$$

    c.　the Murray-Landis variable-grid, space transforma-
tion [10], for the initial conditions:

$$C = C_{\alpha o} = 0 \qquad \frac{\ell}{2} < x \le \frac{L}{2}$$

$$C = C_{\beta o} = 1.0 \qquad 0 \le x < \frac{\ell}{2}$$

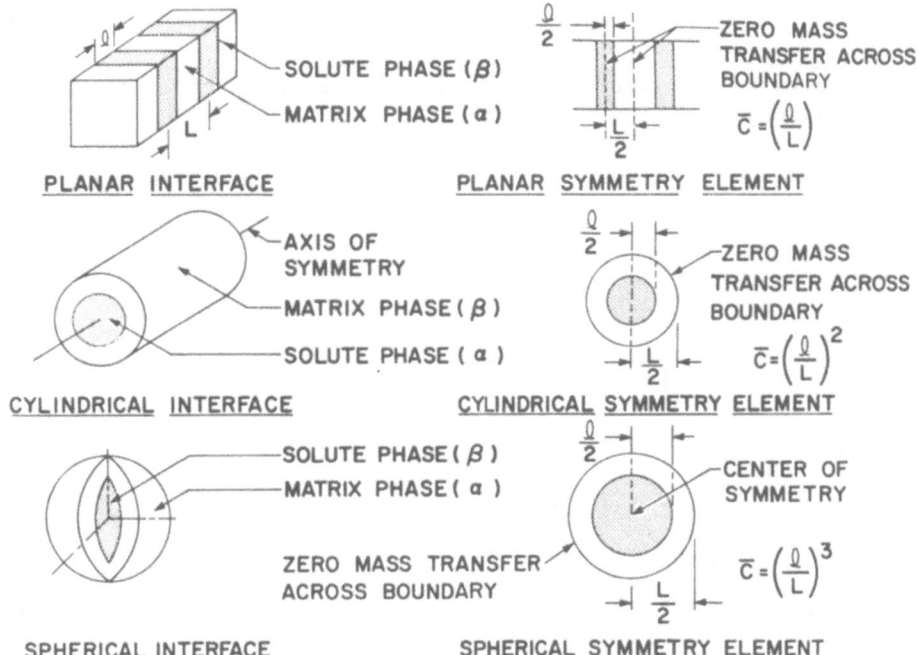

Figure 8. Geometric models showing symmetry elements for planar, cylindrical, and spherical interfaces. (Tanzilli and Heckel [9]; courtesy fo the Transactions of The Metallurgical Society of AIME).

the interface conditions:

$$C = C_{\alpha\beta} \qquad x = \frac{\xi^+}{2} \qquad t > 0$$

$$C = C_{\beta\alpha} \qquad x = \frac{\xi^-}{2} \qquad t > 0$$

the boundary conditions:

$$\frac{\partial C^\beta}{\partial x} = 0 \qquad x = \frac{L}{2} \qquad t \geq 0$$

$$\frac{\partial C^\alpha}{\partial x} = 0 \qquad x = 0 \qquad t \geq 0$$

and
$$\bar{C} = \left(\frac{\ell}{L}\right)^m$$

The setup of the two-phase, moving interface problem is given in Figure 9.

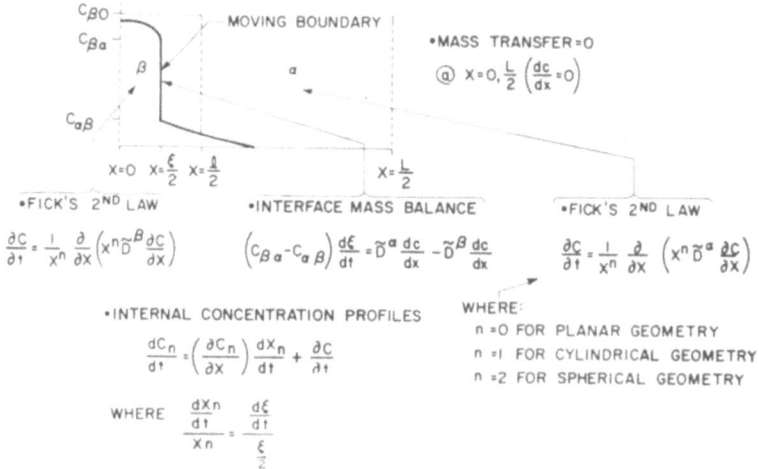

Figure 9. Analytical model for the two-phase, moving inter-face problem for a typical symmetry element. The concentration profiles in each phase are described by Fick's second law. The interface velocity at $x = \xi/2$ is described by means of a mass balance between phases. Also shown is the Murray-Landis variable-grid, space transformation which describes the time rate of change of concentration at an internal point, whose location is always a constant percentage of the instantaneous phase thickness.

The increased number of input parameters in the two-phase homogenization problem results in a wide variety of possible homogenization phenomena. Figure 10 shows that initially the $\alpha:\beta$ interface may move in either direction (or remain stationary), depending on the relative values of the fluxes (J) at the interface. When the homogenization becomes "finite" (i.e., the concentrations at $x = 0$ and/or $x = L/2$ depart from their initial concentrations) the interface moves toward $x = 0$,

in agreement with the restriction that $\bar{C} < C_{\alpha\beta}$. Figure 10
shows that the intermediate stages of homogenization gener-
ally take place with a minimal concentration gradient in
the β phase. The final stage of homogenization occurs following the depletion of the β phase and is analogous to
homogenization in a one-phase system.

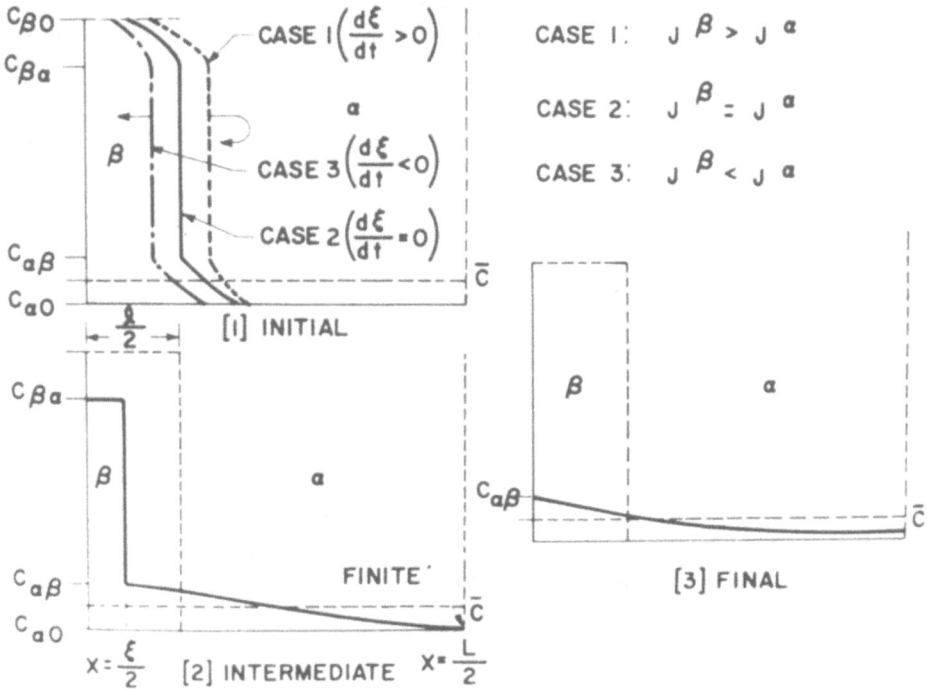

Figure 10. Schematic concentration-distance profiles illustrating the stages of homogenization for a two-phase composite
whose mean composition lies in a single-phase region. Initially, the α:β interface may move in either direction (Cases
1 and 3) or remain stationary (Case 2) depending on the relative values of the fluxes (J) at the interface. (Tanzilli
and Heckel [9]; courtesy of the Transactions of The Metallurgical Society of AIME).

Tanzilli and Heckel [9] evaluated the approach to homogeneity by two parameters:
  a. ξ/ℓ, the normalized size of the β phase,

  b. $(C/\bar{C})_{x\,=\,L/2}$, the normalized departure from the mean composition at the location $x = L/2$.

Their calculations indicated that the rate of solution of the β phase ($\xi/\ell \rightarrow 0$) is primarily dependent upon D^α, ℓ, and $C_{\alpha\beta}$, and geometry (whether planar, cylindrical or spherical) and relatively independent of D^β, $\bar{C}$, and $C_{\beta\alpha}$. These results are shown in Figures 11, 12, and 13, where:

$$\phi = \frac{D^\beta}{D^\alpha} \tag{11}$$

$$\delta = C_{\beta\alpha} - C_{\alpha\beta} \tag{12}$$

and

$$\lambda = \frac{C_{\beta o} - C_{\beta\alpha}}{C_{\alpha\beta} - C_{\alpha o}} \tag{13}$$

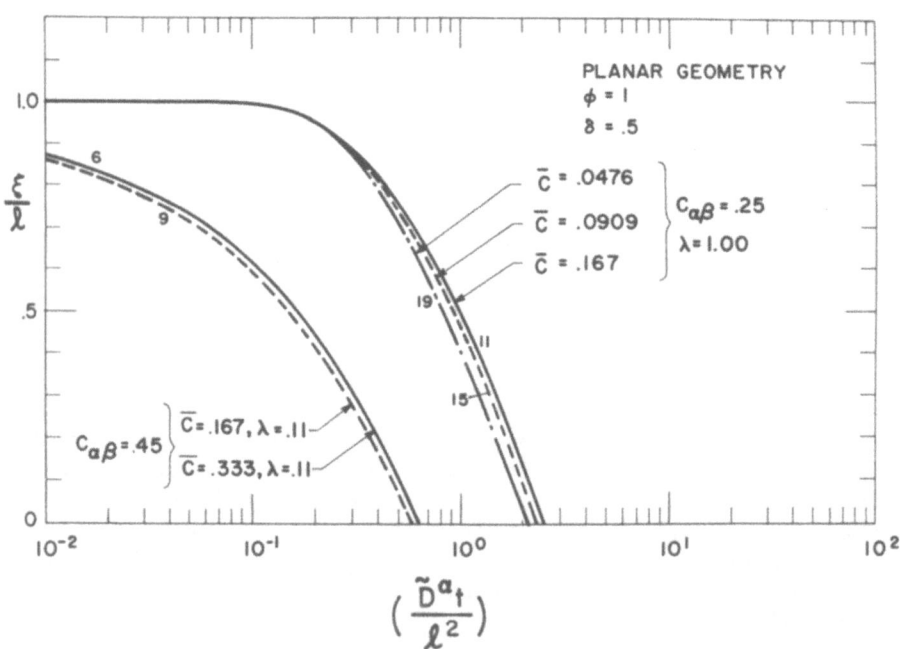

Figure 11. Variation in solution behavior of the unstable β phase as a function of mean composition, $\bar{C}$, (minor effect) and solubility, $C_{\alpha\beta}$ (major effect). (Tanzilli and Heckel [9]; courtesy of the Transactions of The Metallurgical Society of AIME.)

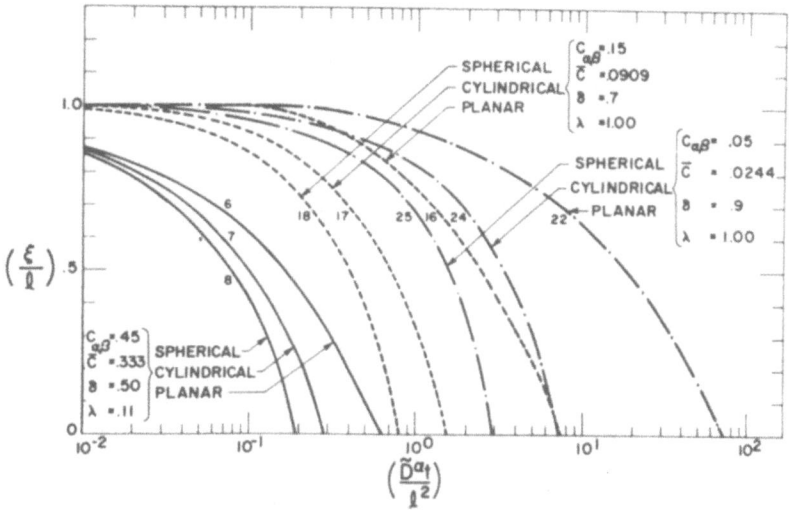

Figure 12. Variation of the solution behavior of the unstable β phase as a function of geometry and solubility, $C_{\alpha\beta}$. Both parameters are shown to have a marked effect on the homogenization process. (Tanzilli and Heckel [9]; courtesy of the Transactions of The Metallurgical Society of AIME).

It is especially interesting to note that "boundary turnaround" (Case 3 in Figure 10) can occur to a marked extent (Figure 13, curve 23) if the flux in the β phase at the α:β interface is large compared to that in the α phase (in this instance, promoted by the large value of λ; in general, promoted by large φ or λ). Thus, it may be found that the phase to be dissolved may initally maintain its size (Figure 13, curves 1 and 22) or may grow (Figure 13, curves 4 and 23) prior to dissolving.

The rate of approach to homogeneity as measured by $(C/\bar{C})_{x = L/2}$ was shown [9] to be primarily a function of D^{α}, ℓ, $\bar{C}$, and geometry and relatively independent of D^{β}, $C_{\alpha\beta}$, and $C_{\beta\alpha}$. These results are given in Figure 14 and provide the basis for the evaluation of true homogeneity, i.e., negligible gradients in the concentric-sphere composite.

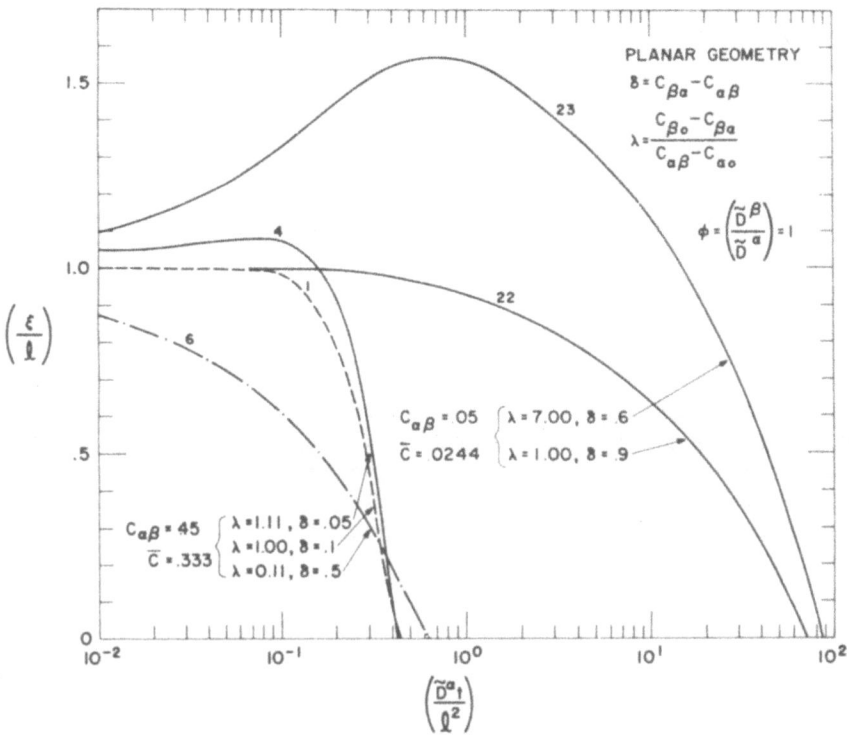

Figure 13. The effect of variations in the β:α concentration range ratio, λ, on interface location is shown for $C_{\alpha\beta}$ values of 0.45 and 0.05. In both instances the spread in the dimensionless time parameter at β depletion, ξ/ℓ = 0, due to λ variations is small compared to the spread due to $C_{\alpha\beta}$. (Tanzilli and Heckel [9]; courtesy of the Transactions of The Metallurgical Society of AIME).

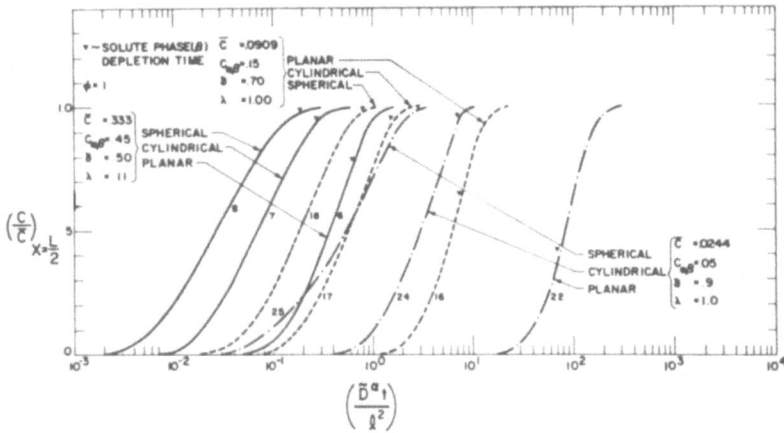

Figure 14. Normalized concentration at the α-phase symmetry boundary, x = L/2, as a function of the dimensionless time parameter, $D^\alpha t/\ell^2$, showing both interface geometry and mean composition, $\bar{C}$, as significant parameters. (Tanzilli and Heckel [9]; courtesy of the Transactions of The Metallurgical Society of AIME).

The evaluation of the major homogenization process parameters in the two-phase system provided the basis for the nomograph in Figure 15 [11]. All (ξ/ℓ) and $(C/\bar{C})_{x = L/2}$ curves obtained over a wide range of input parameters condensed to the relatively narrow bands shown in Figure 15 when plotted as functions of $(D^\alpha t/\ell^2)C_{\alpha\beta}^y$ and $(D^\alpha t/\ell^2)\bar{C}^y$, respectively, where the geometry constant, y, has values of 2.10, 1.40, and 1.10 for planar, cylindrical, and spherical geometries, respectively. The ξ/ℓ curves generally ranked within the band according to the parameter δ/λ; $\delta/\lambda \cong 0.1$ represents the upper portion of the band and $\delta/\lambda \cong 1.0$ represents the lower portion of the band. No ranking in the ξ/ℓ band was observed for various geometries; the $(C/\bar{C})_{x = L/2}$ curves separated into sub-bands for each geometry. The nomograph at the top of Figure 15 provides a graphical calculation of the abscissa value for the (ξ/ℓ) and $(C/\bar{C})_{x = L/2}$ plots. The sample calculation (dashed lines) is shown for $\bar{C} = 0.01$, $C_{\alpha\beta} = 0.04$, spherical geometry (y = 1.10), and $D^\alpha t/\ell^2 = 2.0$. Figure 15, therefore, provides the complete graphical representation of the two-phase, concentric-sphere, homogenization model and is the counterpart to Figures 5 and 6 for the one-phase model.

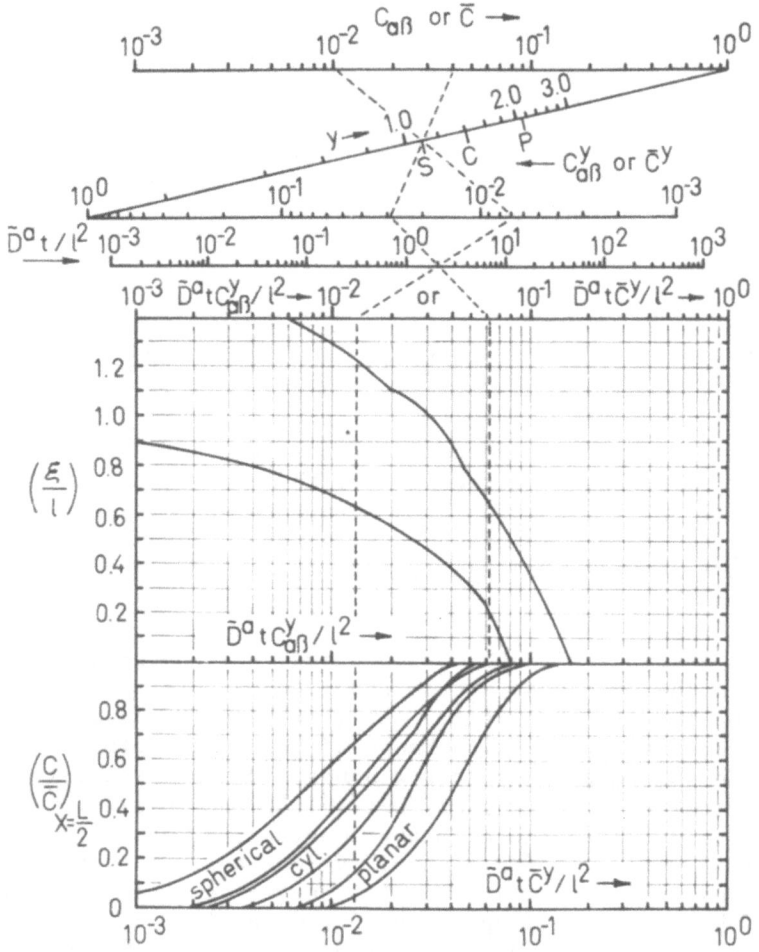

Figure 15. Nomograph or the approximation of ξ/ℓ and $(C/\bar{C})_{x \,=\, L/2}$ as a function of the major process variables for the two-phase, homogenization model. The upper portion provides a nomograph to facilitate the calculation of $(D^\alpha t/\ell^2)C_{\alpha\beta}^y$ for the ξ/ℓ approximation or $(D^\alpha t/\ell^2)\bar{C}^y$ for the $(C/\bar{C})_{x \,=\, L/2}$ approximation. Selection of the appropriate concentration variable ($C_{\alpha\beta}$ for ξ/ℓ; $\bar{C}$ for $(C/\bar{C})_{x \,=\, L/2}$), geometry parameter, y, (planar, 2.10; cylindrical, 1.40; spherical, 1.10), and reduced time, $D^\alpha t/\ell^2$, permits a graphical calculation of the appropriate abscissa value. (Tanzilli and Heckel (11); courtesy of the Transactions of The Metallurgical Society of AIME).

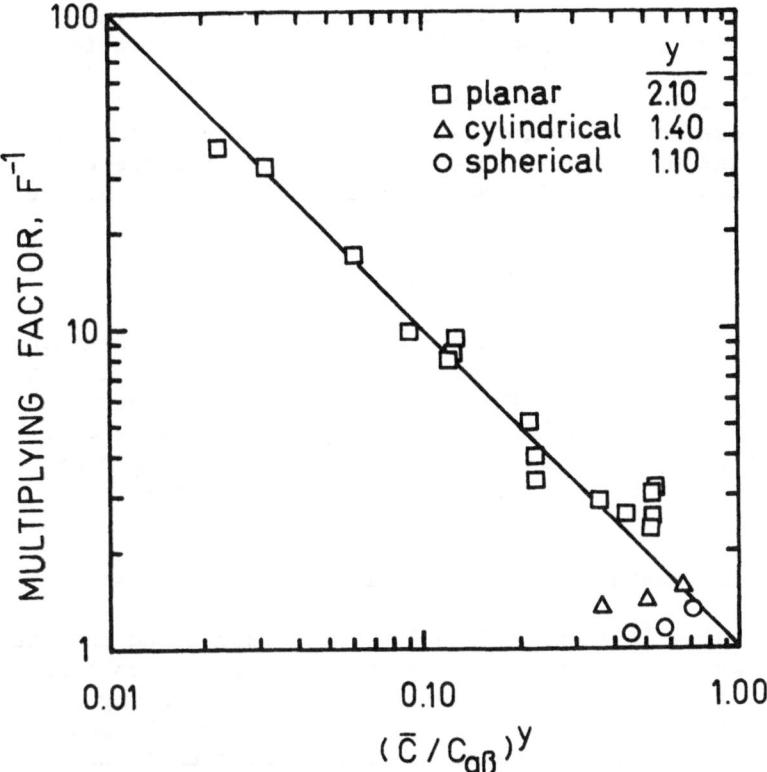

Figure 16. Correlation between the ratio of the time for
essentially complete homogenization to the time for depletion
of the unstable phase (F^{-1}) and $(\bar{C}/C_{\alpha\beta})^y$. (Tanzilli and
Heckel [11]; courtesy of the Transactions of The Metallur-
gical society of AIME).

Figure 15 also suggests a simplification that may be used
in considering homogenization in two-phase systems [11]. Essentially complete homogenization is generally a more difficult
condition to determine experimentally than the time to cause
depletion of the β phase ($\xi/\ell = 0$). However, it may be seen
from Figure 15 that:

$$\xi/\ell = 0 \quad \text{at} \quad \frac{D^\alpha t}{\ell^2} \cdot C_{\alpha\beta}^y \cong 0.10 \tag{14}$$

$$\text{and} \quad (C/\bar{C})_{x \,=\, L/2} \text{ at} \quad \frac{D^\alpha t}{\ell^2} \cdot \bar{C}^y \cong 0.10 \tag{15}$$

Therefore, the ratio of the time for depletion of the β phase
and the time to get essentially complete homogenization is
approximately $(C_{\alpha\beta}/\bar{C})^y$. Thus, an experimentally-determined
time for depletion when multiplied by $(C_{\alpha\beta}/\bar{C})^y$ will give the
approximate total homogenization time. The correlation
between the multiplying factor and $(\bar{C}/C_{\alpha\beta})^y$ for the computer
calculations of Tanzilli and Heckel[9] is given in Figure
16 [11].

C. Three-Phase Binary Systems

A comprehensive treatment of homogenization phenomena
in the three-phase system as described in Figure 2 is currently not available. The concentric-sphere geometry shown
in Figure 3 to be applicable to two-phase systems should
permit the adptation of previously developed methods of
analysis [9]. Such a model is being prepared [12].

D. General Multiphase-Multicomponent Models

Kirkaldy [13] and Kidson [14] have formalized the treatment of multiphase and multicomponent diffusion problems.
However, the development of general models including all
possible input parameters for planar, cylindrical, and
spherical geometries appears to be a mammoth undertaking.
A better approach would seem to be to consider in detail
binary one-, two-, and three-phase and ternary one- and
two-phase models. The insight into systems of greater
complexity may be possible by "extrapolating" to achieve
a semi-quantitative understanding.

VI.   EXPERIMENTAL TECHNIQUES

A. Various Techniques Applicable to
Homogenization

A wide variety of experimental techniques has been used
by previous authors to study the homogenization of compacted
blends of powders. Examples of studies that demonstrate
these techniques include electrical resistivity [15-18],
dilatometry [15, 19-21], differential thermal analysis
[2, 23, 24], microscopy [25, 26], x-ray compositional line
broadening [1, 8, 21, 27, 28], electron microprobe measurements [8, 29, 30], quantitative metallography techniques [31],

and x-ray techniques for determination of amounts of phases
(29). Although all of these techniques have been proven to
provide measures of the homogenization process, the more
recently applied techniques such as x-ray compositional line
broadening, electron microprobe measurements, and quantitative determination of amounts of phases appear to provide
the most direct, quantitative information concerning the
state of heterogeneity in a powder compact. In addition,
these techniques provide the most direct information necessary for comparison of the previously discussed models to
actual powder compacts. The remainder of this paper will
be concerned with the application of these quantitative
techniques to compact homogenization studies and the interpretation of the data provided by them.

B. One-Phase Systems

Electron Microprobe Techniques. The capability of the
electron microprobe to determine the composition of an area
in the microstructure of a few microns in diameter makes
this research tool directly applicable to quantitative homogenization studies. Since the variation of concentration
with distance is the principal structural feature to be
studied in one-phase homogenization homogenization, microprobe scans across the microstructure provide a direct measurement of the degree of heterogeneity.

White [30] has used the electron microprobe to follow
the homogenization of hot-pressed powder blends having a
mean composition of W - 10Mo - 10Cb.* Figures 17 and 18 [30]
show the effect of temperature on the heterogeneity and
microstructure of this powder fabricated alloy. Not only do
the microprobe scans show the increase in homogeneity with
increasing temperature, but also the relatively rapid leveling of the molybdenum gradients due to the fine particle
size of this powder.

Electron microprobe scans of copper-nickel ($\bar{C}_{Ni}$ = 0.52)
obtained in a previous investigation [8] are presented in
Figure 19. Three scans at different specimen locations are
shown for each homogenization treatment. The nickel concentration ranges between C_{Ni} = 0 at the background level

---

* Weight percent

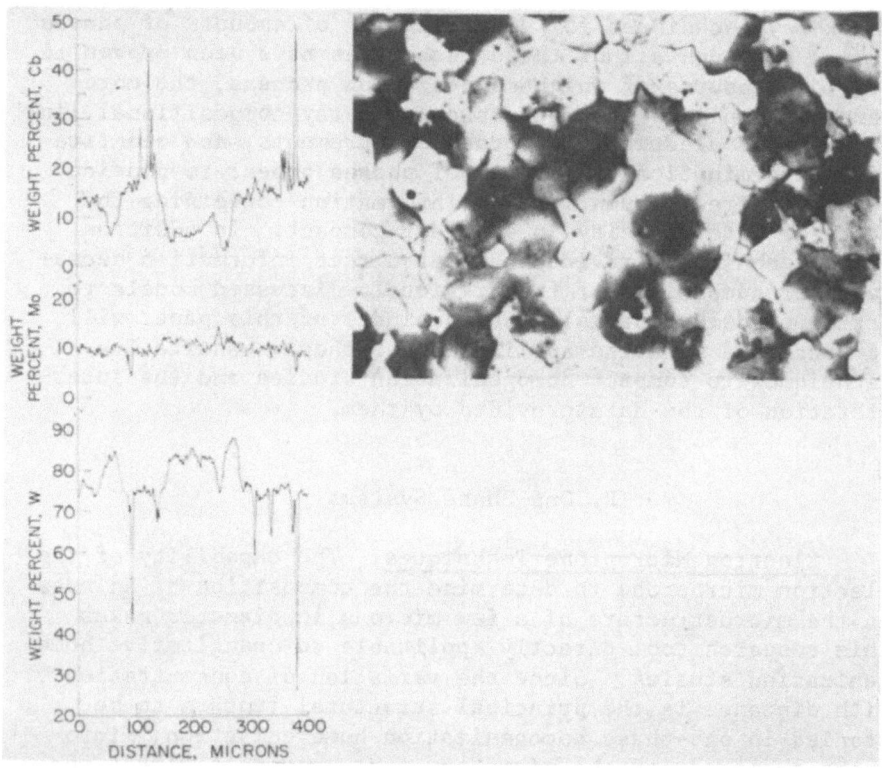

Figure 17. Correlation between heterogeneities revealed by non-uniform etching and electron microprobe scans in a hot-pressed (2100° C, 30 min., 6000 psi.) blend of tungsten, molybdenum, and columbium powders of mean composition. W-10^W/oMo-10^W/oCb. Particle sizes: W and Mo, 1μ; Cb, 10 to 40μ. (White [30]; courtesy of the Transactions of ASM).

to C_{Ni} = 1.0 at the pure nickel (standard) level. Although the nickel concentration is approximately proportional to intensity, alloy standard and/or calculated corrections can be used to obtain greater precision (8, 22). The abrupt fluctuations in the sample current shown on Figure 19 are useful for defining the location of porosity in the path of the scan. Scanning speeds should generally be chosen to provide resolution of the maxima and minima in the composition-distance profiles. In Figure 19 the 100 micron scanning distances were approximately twice the powder particle diameter.

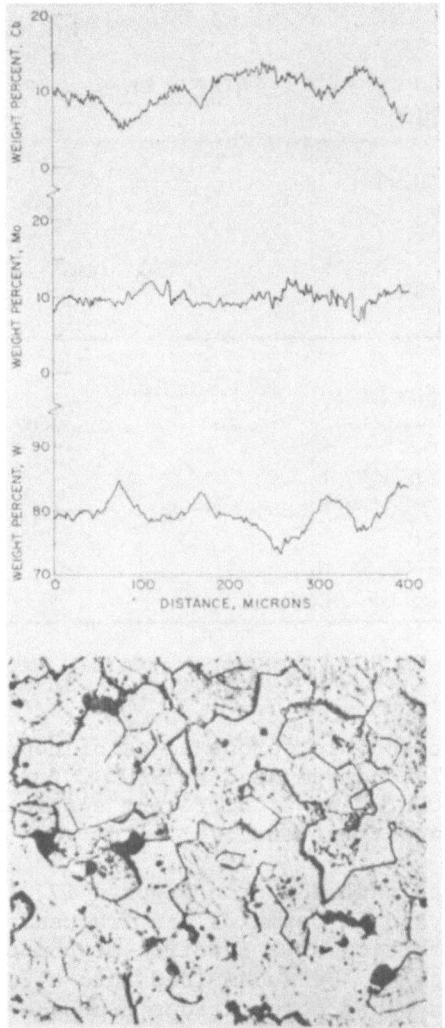

Figure 18. Correlation between the relatively homogeneous structure (compared to Figure 17) revealed by etching and electron microprobe scans after high-temperature treatment (2600° C, 20 min.) of a compacted blend of tungsten, molybdenum, and columbium powders of mean composition. W-10^W/oMo-10^W/oCb. Particle sizes: W and Mo, 1μ; Cb, 10 to 40μ. (White [30]: courtesy of the Transactions of ASM).

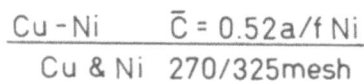

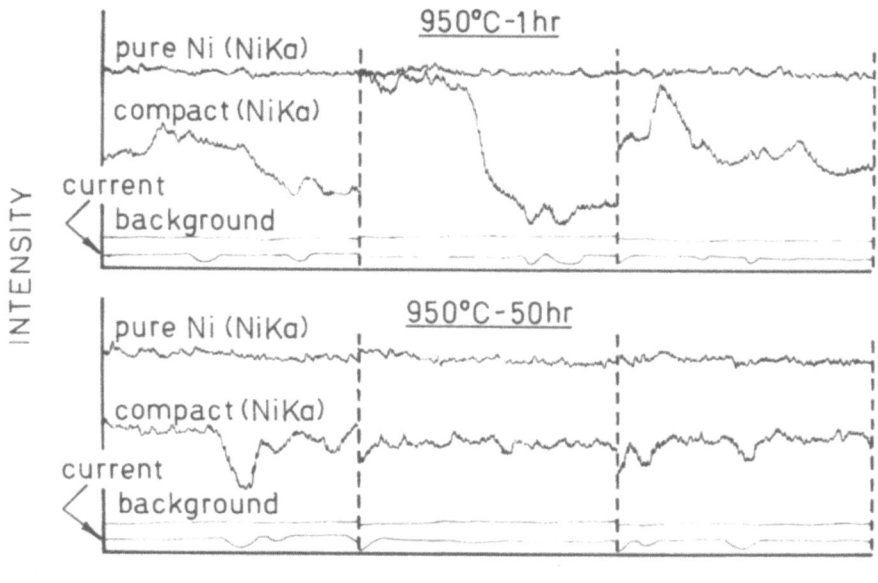

DISTANCE

Figure 19. Electron microprobe scans of copper-nickel compacts ($\bar{C}_{Ni}$ = 0.52) homogenized at 950° C for 1 hour and 50
hours.   Three 100μ scans at different specimen locations
are shown for each treatment.

Data such as that shown in Figures 17, 18, and 19 may
be readily analyzed in terms of the frequency of occurrence
of a given composition as a function of composition. A
given scan length may be subdivided into a large number of
small increments whose compositions can be determined from
the microprobe output. The narrowness of the frequency-
composition plot obtained is a measure of the homogeneity
of the compact and is directly related to the value of
($C_{A\ max.}$ - $C_{A\ min.}$) which may be obtained from the concentric-
sphere model (Figure 5).

Duwez and Jordon[3] have noted that concentrations in
mathematical models having the lowest concentration gradients
should correspond to concentrations having high frequencies
of occurrence in powder compacts and vice versa. Since zero

gradients at both $r/b = 0$ ($C_{A\ max.}$) and $r/b = 1.0$ ($C_{A\ min.}$)
are boundary conditions for the concentric-sphere model
(Figure 4), it should be expected that frequency-composition
plots should exhibit two maxima. These maxima should merge
during the later stages of homogenization.

Rudman [27] has analyzed frequency-composition plots
obtained from experimental data and related them to effective concentration-distance profiles which were analogous
to concentration-distance profiles developed from mathematical models. Since $(dC/dx)^{-1}$, the reciprocal of the concentration gradient in a mathematical model, is proportional
to the frequency of occurrence of the concentration at that
gradient, the normalized integral of the frequency-composition plot:

$$\int_0^C N(C)\,dC \bigg/ \int_0^1 N(C)\,dC \tag{16}$$

plotted as a function of concentration should be the effective concentration-distance profile for a given set of experimental data ($N(C)$ is the frequency of occurrence of concentration C). Rudman's method has been successfully applied
to microprobe data [8] even though it was originally developed for x-ray compositional line broadening data [27].

X-Ray Compositional Line Broadening. Rudman [27] has
analyzed the compositional heterogeneities in solid solutions in terms of the broadening of x-ray diffraction peaks.
This form of line broadening is a function of the variation
of lattice parameter with composition; a range of concentration, ΔC, gives a range of lattice parameter, Δa, which, in
turn, gives a range of diffraction angles, $\Delta 2\theta$. This technique is generally applicable to situations where the lattice parameter varies more than about 0.03Å over the range
of composition heterogeneity to be studied. Rudman's studies
on the homogenization of copper-nickel powder compacts have
indicated that the diffracted intensity vs. 2θ profiles for
a given hkl peak can be corrected to provide frequency vs.
composition plots ($N(C)$ vs. C). This data was used in conjunction with Equation 16 and the method described previously
to obtain effective concentration-distance profiles for
comparison to the concentric-sphere model [1, 21].

Leber and Hehemann [28] have used the compositional
line broadening technique of Rudman [27] to study the homogenization of powder-fabricated Cb-15W-5Mo-1Zr * (F48).
The frequency-composition plots obtained for various temperatures of homogenization are shown in Figure 20, where the
composition range is seen to become narrower (increased homogeneity) with increasing temperature. These line broadening
results are in agreement with the metallographic results of
Wolff [26] in Figure 21. Both Figures 20 and 21 indicate
heterogeneities that exist after homogenization at 2300° C.

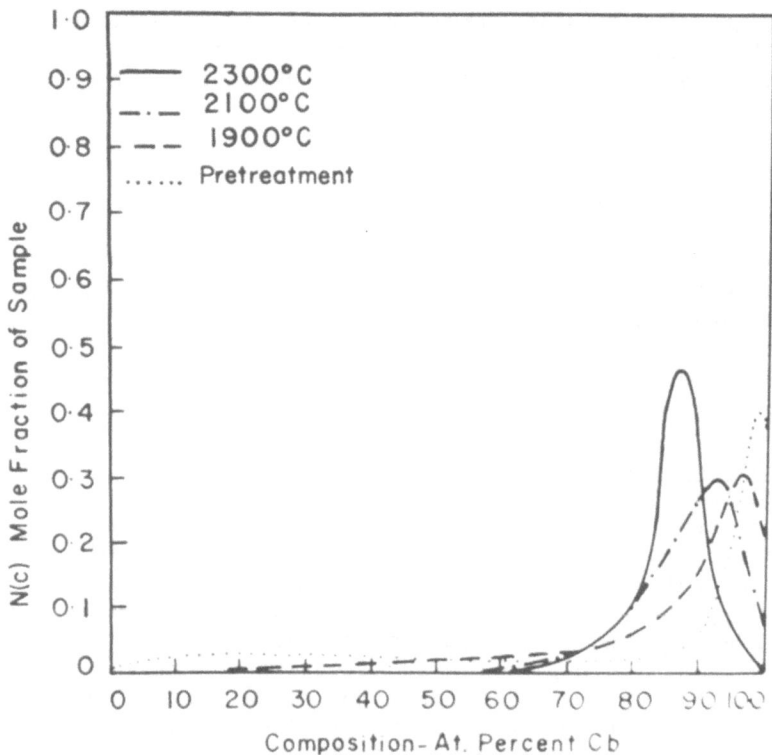

Figure 20. Composition distributions determined by x-ray
compositional line broadening after various 1 hour homogenization treatments of compacted powder blends having the mean
composition Cb-15^W/oW-5^W/oMo-1^W/oZr. (Leber and Hehemann
[28]; courtesy of the Transactions of The Metallurgical
Society of AIME).

---

* Weight percent

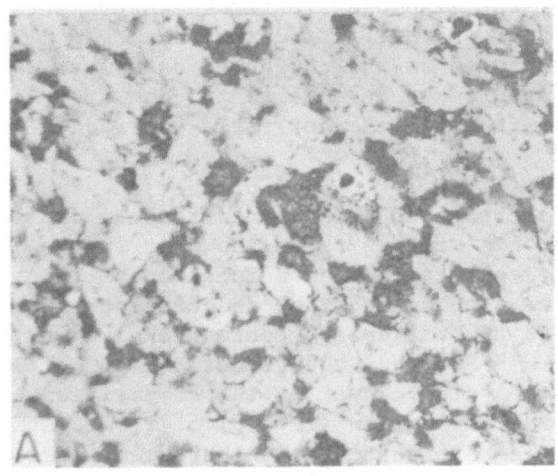

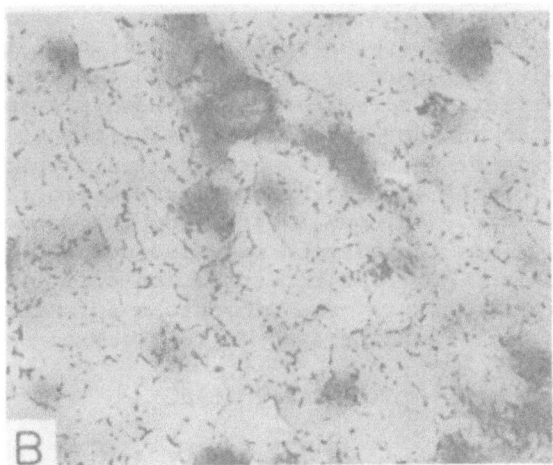

Figure 21. Metallographic evidence of the homogenization of compacted powder blends of mean composition Cb-15^w/oW-5^w/oMo-1^w/oZr for comparison to the x-ray compositional line-broadening data shown in Figure 20. A - 1900° C, 1 hour; B - 2300° C, 1 hour. Dark areas are tungsten-rich; light areas are molybdenum-rich. (Wolff [26]; courtesy of the Transactions of ASM).

Heckel [8] has used the Rudman analysis [27] to study
the homogenization of copper-nickel compacts over ranges
of time, temperature, and composition. The Fourier method
of Stokes [33] was also used to remove instrumental broadening effects from the diffracted intensity vs. 2θ data; the
Fourier coefficients of the corrected data were obtained by
the computer program of De Angelis and Schwartz [34] and
were used to synthesize the corrected diffracted intensity
vs. 2θ curve.

The 311 diffraction peaks (uncorrected) from a series
of samples in this study [8] are shown in Figure 22. The
progress of homogenization (as a function of time and temperature) can be seen to occur by the development of compositions between the extremes of pure nickel (2θ $\cong$ 93°) and
pure copper (2θ $\cong$ 90°) and the gradual merger of the two
main peaks into a single peak at the 2θ position corresponding to the mean composition (2θ $\cong$ 91.5°). The sharpening
of the single peak to that of the instrumental broadening
width shown by the pure nickel and pure copper peaks (upper
left part of Figure 22) represents the final stage of homogenization.

The correlation between x-ray compositional line broadening data, electron micrpprobe data, and the concentric-
sphere homogenization model have been shown previously [8].
Figure 23 [8] shows that the concentration-effective penetration analysis of Rudman [27] provides data which agree
with the form of the concentration-distance profiles (plotted
as concentration vs. volume $(r/b)^3$) from the concentric-
sphere model. This correlation was also pointed out by
Fisher and Rudman [1]. The concentration-effective penetration curves for both x-ray and microprobe data ideally begin
at t $=$ 0 as a vertical line at y $=$ $(1-\bar{C})$ where:

$$y \;=\; \int_0^C N(C)dC \;/\; \int_0^1 N(C)dC \qquad\qquad (17)$$

and end at t $= \infty$ as a horizontal line at $\bar{C}$. Intermediate
times are characterized as a sequence of curves varying in
position between these extremes.

Fisher and Rudman [1] have shown that the degree of
interdiffusion, (Equation 5), may be determined from the
concentration-effective penetration curves by:

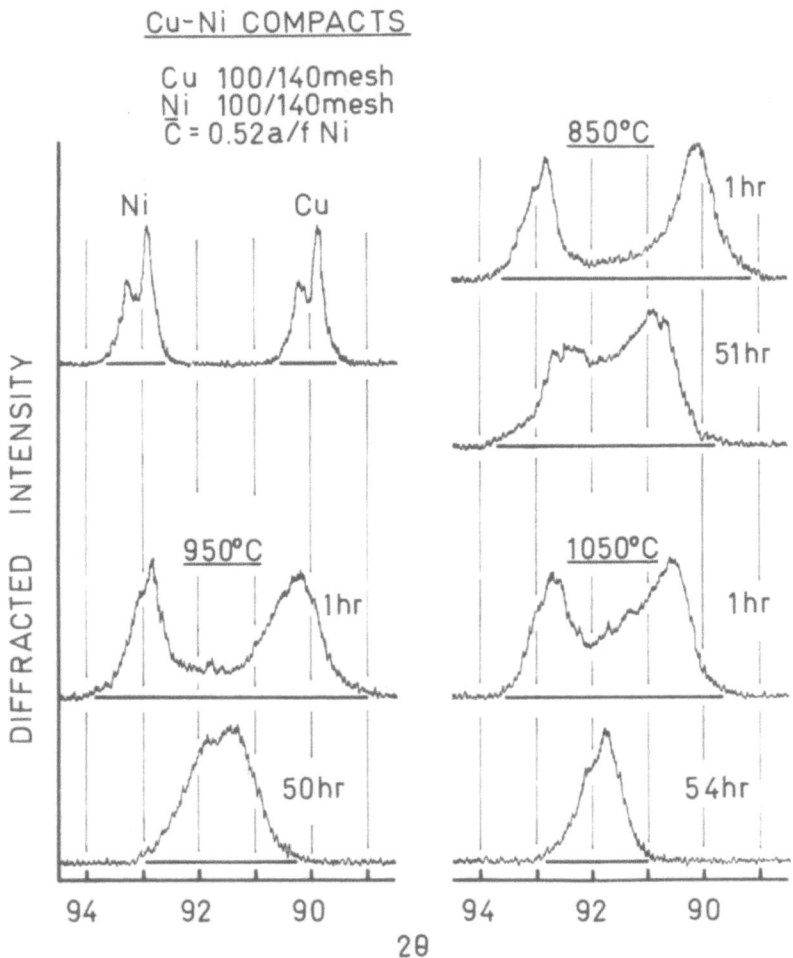

Figure 22.   Uncorrected 311 CuK$_\alpha$ diffraction peaks ($\bar{C}_{Ni}$= 0.52).
The initial condition of the compact is represented by the
sharp diffraction peaks for pure copper and pure nickel in
the upper left portion of the illustration. At a given temperature the progress of homogenization may be followed by
means of the x-ray compositional line broadening associated
with the creation of a range of solid solution compositions
due to interdiffusion. At later times, the broadened peaks
merge as the composition profile approaches the mean alloy
composition.

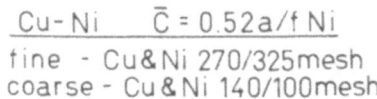

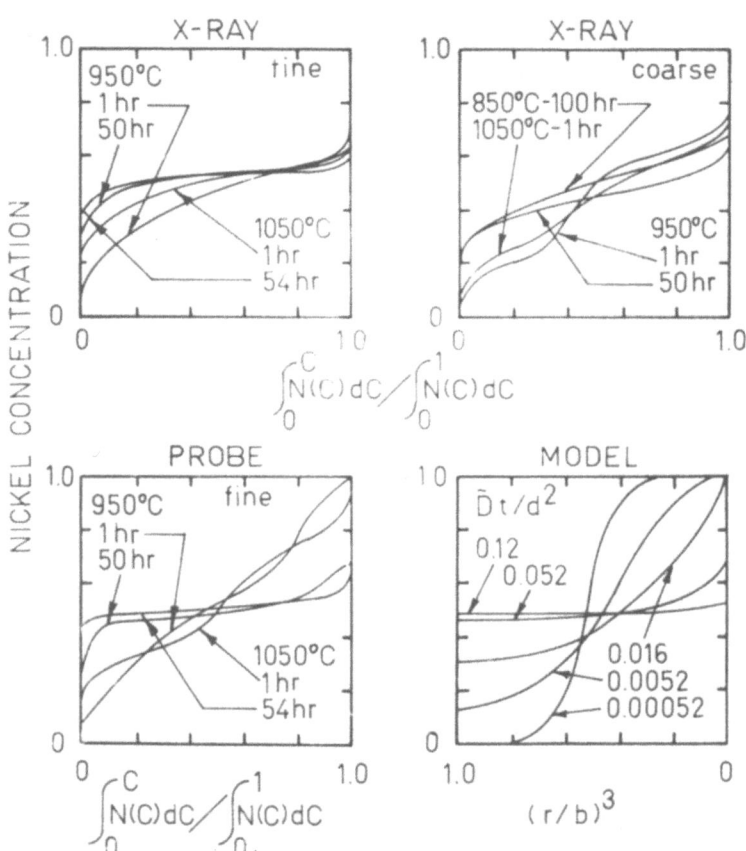

Figure 23. Concentration penetration curves determined from x-ray compositional line broadening, electron microprobe scans and the concentric-sphere model, using the Rudman (27) formalism. Comparison of the effective nickel penetration curves for the fine powder compacts which were deduced from x-ray compositional line broadening and electron microprobe data show good agreement (graphs in left portion of illustration). The effect of coarse initial particle size is seen to retard the progress of homogenization (compare fine and coarse data at 50 hours, 950° C - upper graphs). The experimental data are shown to correspond to the form of curves derived from the concentric-sphere model (lower right).

$$m_t = \int_0^{C_{y=y'}} (y' - y)dC + \int_{C_{y=y'}}^1 (y - y')dC \qquad (18)$$

and
$$m_\infty = \bar{C}y' + (1 - y')(1 - \bar{C}) \qquad (19)$$

where the Matano interface, $y = y'$, is defined by:

$$\int_0^{C_{y=y'}} (y' - y)dC = \int_{C_{y=y'}}^1 (y - y')dC \qquad (19)$$

The correlation between F values obtained from the concentration-effective penetration curves (x-ray and microprobe data) and F values obtained from the concentric-sphere model in a study by Heckel [8] are shown in Figure 24. The experimental data points for each mean composition represent various temperatures and particle sizes; the normalization of the experimental data to a single line by the parameter Dt/d_A^2 verifies the general applicability of the concentric-sphere analysis. The departure of the data from the concentric-sphere model at large values of τ was concluded to result from non-ideal interparticle mixing and was evaluated in terms of an effective nickel particle size which increased with increasing homogeneity (Figure 25).

The range of nickel concentrations existing in the copper-nickel compacts was also evaluated in this study [8] and compared to the predictions of the concentric-sphere model (Figure 26). Again, excellent correlation was obtained between the x-ray and microprobe techniques, the data for various particle sizes and temperatures were successfully correlated by the parameter, and the rate of homogenization departed from the prediction of the concentric-sphere model with increasing homogeneity.

C. Two-Phase Systems

Quantitative Metallography. As mentioned previously, the two-phase binary homogenization process exhibits a moving heterophase interface during the early stages of homogenization (Figure 2) which enables one to follow these early stages of diffusion annealing by quantitative metallographic techniques. For binary alloys where the differences in the density:molecular weight ratios are small, ξ/ℓ, the

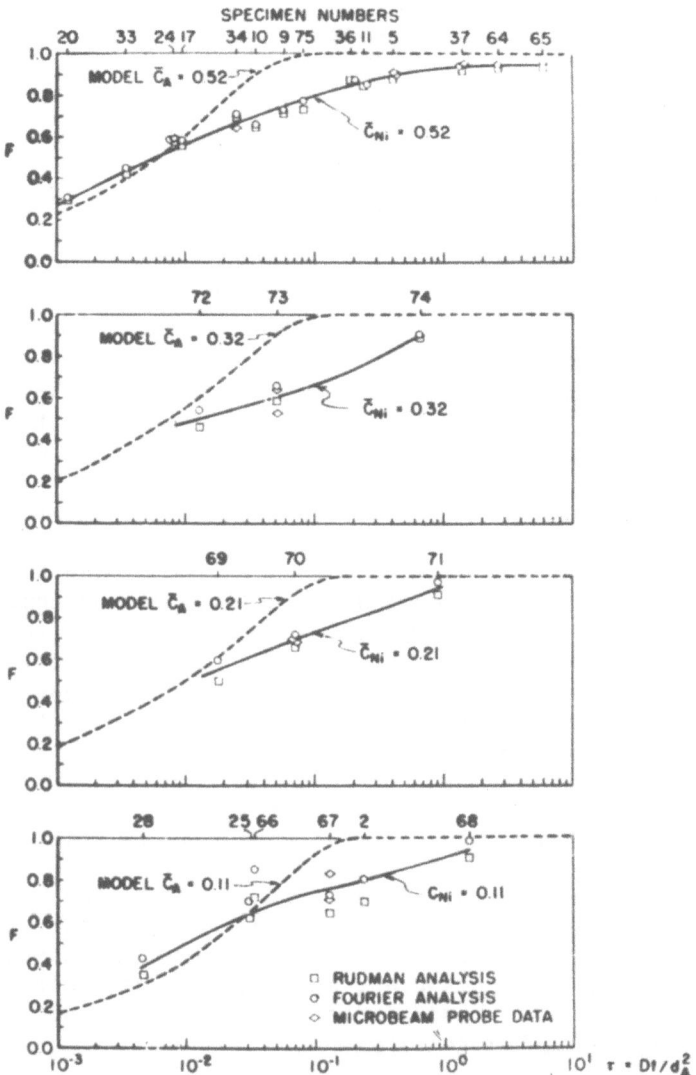

Figure 24. Comparison of experimental (x-ray compositional
line broadening and electron microprobe) and predicted (concentric-sphere model) values of the degree of interdiffusion,
F, as a function of Dt/d_A^2 for copper-nickel compacts of mean
composition, $\bar{C}_{Ni}$, of 0.11, 0.21, 0.32, and 0.52. Both experimental techniques agree. In addition, experimental data for
various particle sizes of nickel, d_A, and various temperatures
(various D's) are normalized to a single curve for each mean
composition by the parameter Dt/d_A^2, in agreement with the
model. (Heckel [8]; courtesy of the Transactions of ASM).

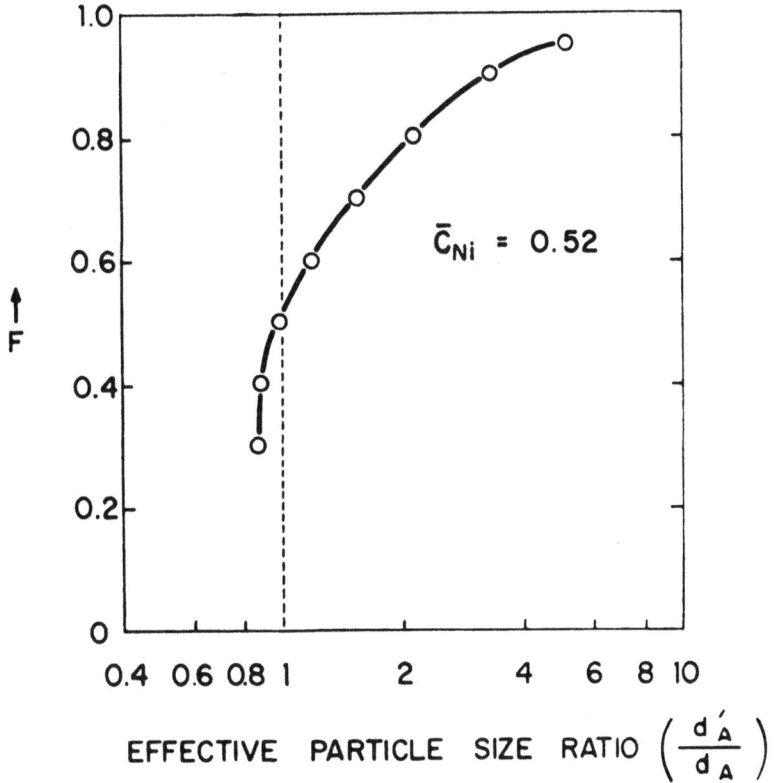

Figure 25. Variation of the effective particle size ratio
with the degree of homogenization, F, for compacts of
blended copper and nickel powders ($\bar{C}_{Ni}$ = 0.52). The effective particle size was determined from the discrepancies
between the predicted and experimental curves in Figure 24
and is indicative of the effects of non-ideal mixing of
powder particles in the as-pressed compact. (Heckel [8];
courtesy of the Transactions of ASM).

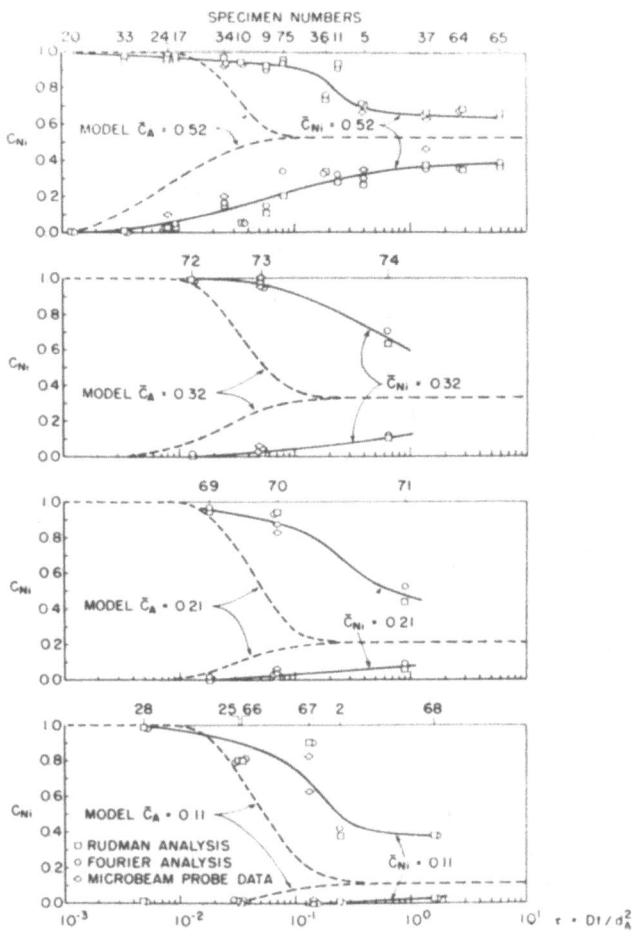

Figure 26. Comparison of experimental (x-ray compositional line broadening and electron microprobe) and predicted (concentric-sphere model) values of the range of composition ($C_{A \text{ max.}}$ and $C_{A \text{ min.}}$) as a function of Dt/d_A^2 for copper-nickel compacts of mean compositions, $\bar{C}_{Ni}$, of 0.11, 0.21, 0.32, and 0.52. Both experimental techniques agree. In addition, experimental data for various particle sizes of nickel, d_A, and various temperatures (various D's) are normalized to a single curve for each mean composition by the parameter dt/d_A^2, in addition with the model. (Heckel [8]; courtesy of the Transactions of ASM).

normalized size of the unstable β phase may be determined
by the relationship:

$$\frac{\xi}{\ell} \cong \left(\frac{V_v}{\bar{C}}\right)^{\frac{1}{3}} \tag{21}$$

where V_v is the volume fraction of the unstable phase at
some time prior to depletion, and $\bar{C}$, the mean alloy composition (a/f), is approximately equal to V_{v_0}, the initial
volume fraction of the unstable phase.

Hilliard and Cahn [35] have shown that the best method
for volume fraction analysis is a systematic point count
in which a grid is superimposed on a sequence of areas
selected either randomly or systematically from the plane
of polish. The total fraction of grid corners falling on a
particular phase provides an unbiased estimate of the volume
fraction of that phase. Thus, the evaluation of the unstable
β phase may be simply obtained using a well-established
metallographic technique. Equation 21 is most accurate for
powder compacts where the unstable phase particle distribution is narrow. For situations where a broad particle size
range exists, one should use supplemental methods for determining ξ and ℓ for the compact (e.g., DeHoff [36]).

The powder compact data shown in Figure 27 were obtained
using the two-dimensional, random point count method for
determining V_v in Equation 21. Photomicrographs for one
experimental sequence on which volume fraction measurements
were made are shown in Figure 3. The experimental data in
Figure 27 show excellent agreement with the analytical model.
Also shown in Figure 27 are data from finite-boundary diffusion couples where direct thickness measurements of the
unstable phase have been made, again showing excellent
agreement with the analytical model.

X-Ray Compositional Line Broadening and Electron Microprobe Scans. X-ray compositional line broadening may be used
to follow the homogenization stages of a two-phase alloy in
a manner similar to that used for single-phase systems. As
may be deduced from Figure 2, the compositional broadening
associated with the two-phase problem differs from the single-
phase problem only in that the range of compositions contributing to the broadening phenomena is narrower, being between
the solubility limit of the stable phase and its initial pure
(or prealloyed) condition. The uncorrected 311 diffraction

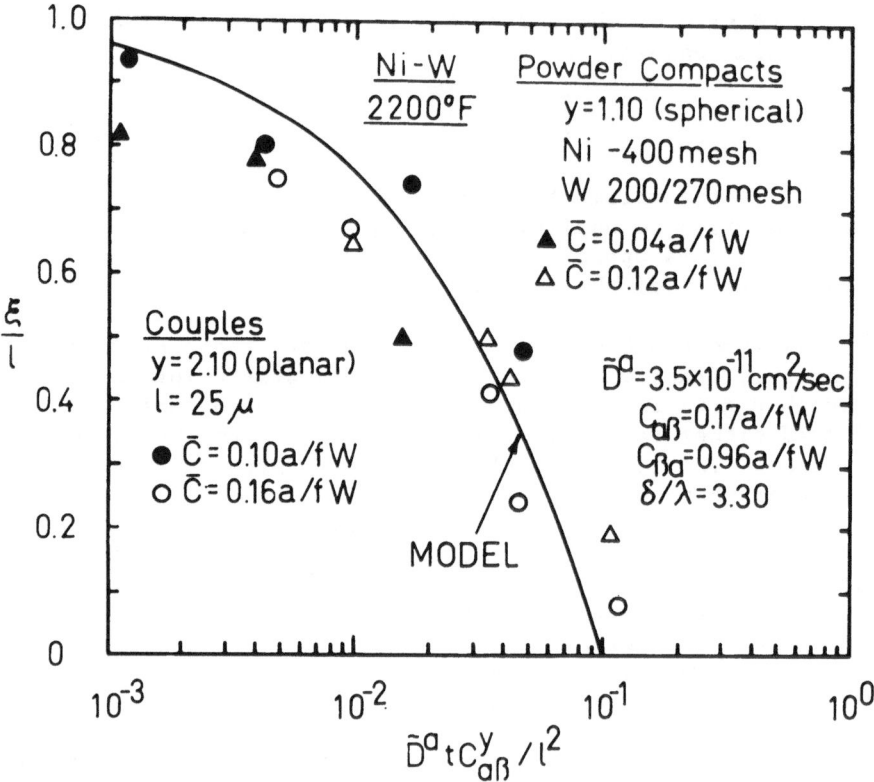

Figure 27. Comparison of experimental and predicted values of ξ/ℓ as a function of $D^\alpha t(C_{\alpha\beta}^y/\ell^2)$ (see Figure 15) for the two-phase, nickel-tungsten system (δ/λ = 3.30). Experimental data for both finite, planar couples and powder compacts are correlated with the results of the mathematical model.

peaks for a series of nickel-tungsten powder compacts are shown in Figure 28. In this case the stable α phase (see Figure 2) is the nickel-rich phase. Initially, the peak of the α phase broadens as the degree of interdiffusion progresses from the pure nickel initial condition. At a later time the peak narrows again as the nickel composition profile approaches the mean alloy composition.

Electron microprobe scans of the compacts and finite boundary couples enumerated in Figure 27 are currently under evaluation [37]. Preliminary evaluation of uncorrected microprobe scans on planar couples shows good agreement with the

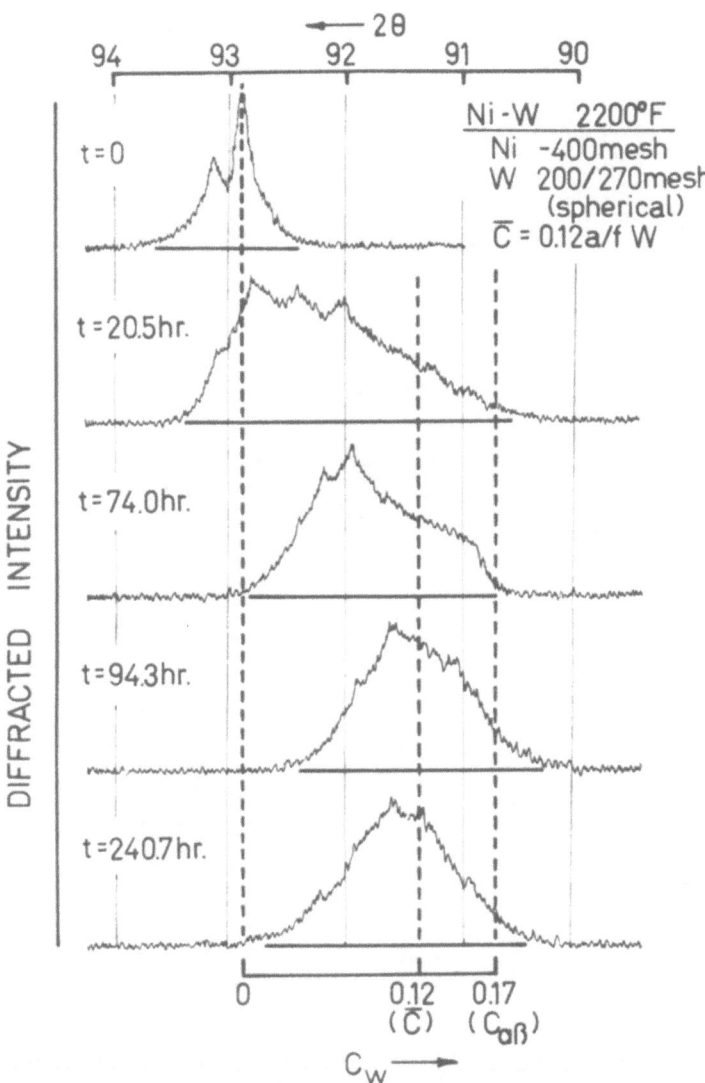

Figure 28. Uncorrected 311 CuK$_\alpha$ diffraction peaks from the
nickel-rich solid solution for nickel-tungsten compacts
($\bar{C}_w$ = 0.12). Since the diffraction angle, 2θ, can be con-
verted to a tungsten concentration through Bragg's Law and
lattice parameter vs. concentration data, the intensity –2θ
data can be reduced to the spectrum in tungsten concentrations in the nickel-rich phase.

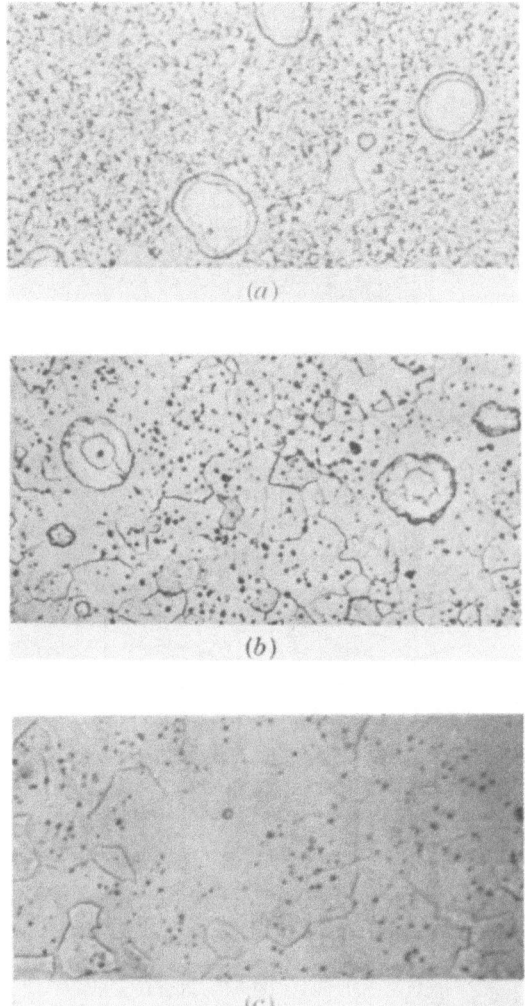

Figure 29. Photomicrographs of compacted blends of tungsten
and rhenium powders (W-5^W/oRe) homogenized at 2325° C for
(a) 1 min., (b) 10 min., (c) 72 min. The various stages
of three-phase homogenization are illustrated: (a) and (b)
rhenium-rich, α-phase particles surrounded by a shell of the
intermediate phase, σ, in a matrix of tungsten-rich β, (c)
one-phase, β, following depletion of α and σ. (Smith and
Hehemann [31]; courtesy of the Transactions of The Metallurgical Society of AIME).

analytical model when $(C/\bar{C})_{x=L/2}$ data are plotted on the lower portion of the nomograph in Figure 15.

D. Three-Phase Systems

Quantitative Metallography. The three-phase system shown in Figure 2 involves two moving heterophase interfaces. Quantitative metallography is, therefore, useful in measuring the progress of homogenization up to the time when only a single phase exists in the compact.

Smith and Hehemann [31] have used quantitative metallography to study the initial stages of homogenization in the three-phase, tungsten-rhenium system (Figures 29, 30, and 31). The three-phase structures shown in Figure 29 (a and b) are schematically illustrated by the early stage of Figure 2 and are quantitatively characterized in Figure 30 ($r'_\alpha > 0$) and Figure 31 ($F_\alpha > 0$). The later stage (two-phase) shown in Figure 2 corresponds to the $r'_\alpha = 0$, $r'_\sigma > 0$ region of Figure 30 and the $F_\alpha = 0$, $F_\sigma > 0$ region of Figure 31. The one-phase structure shown in Figure 29c corresponds to $r'_\sigma = 0$ in Figure 30 and $F_\sigma = 0$ in Figure 31.

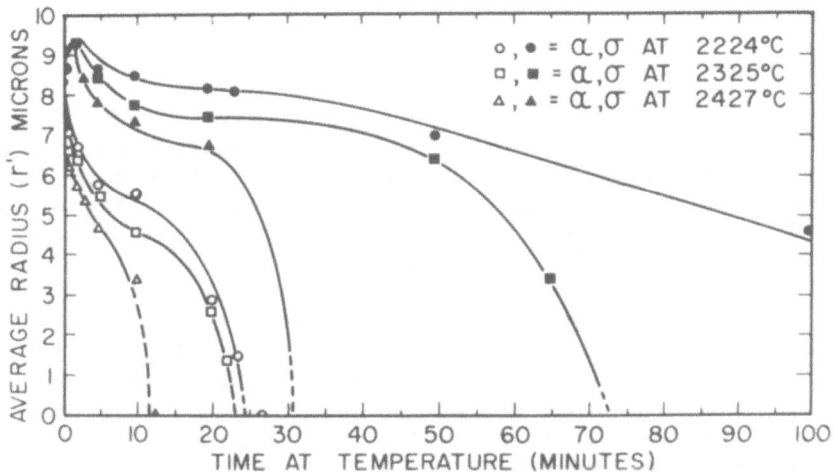

Figure 30. Quantitative metallography measurements of the average radii of the α and σ phases in the three-phase homogenization of tungsten-rhenium compacts shown in Figure 29. (Smith and Hehemann [31]; courtesy of the Transactions of The Metallurgical Society of AIME).

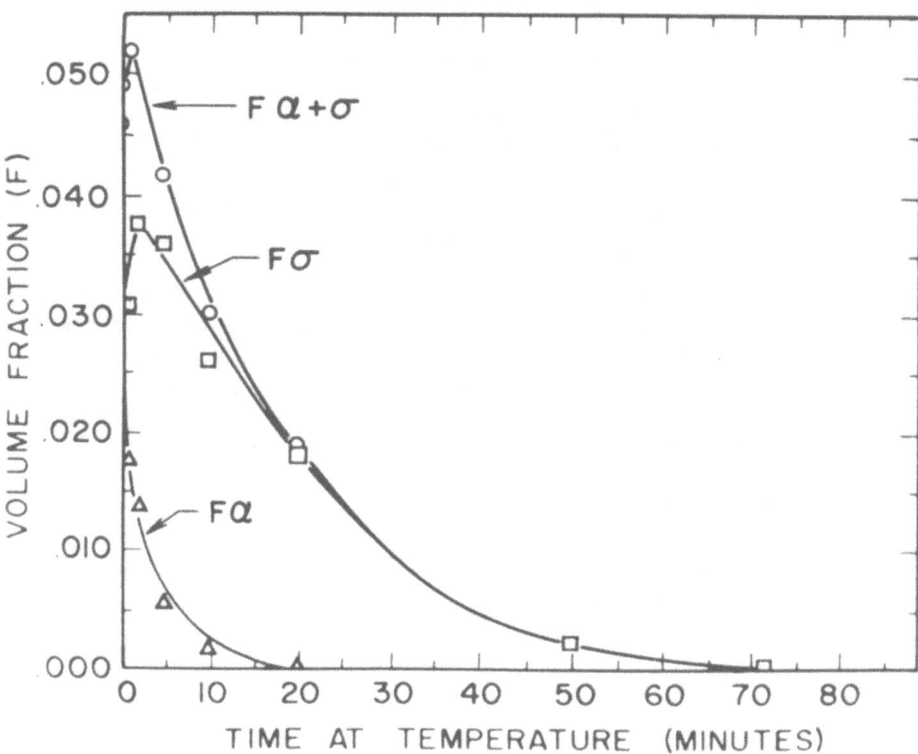

Figure 31.   Quantitative metallography measurements of the
volume fractions of α and σ phases in the three-phase homogenization of tungsten-rhenium compacts shown in Figure 29.
(Smith and Hehemann [31]; courtesy of the Transactions of
the Metallurgical Society of AIME).

The quantitative metallography measurements of mean
phase radius (r_α' and r_σ') and volume fraction (F_α and F_σ) by
Smith and Hehemann[31] provide a quantitative description
of the multiphase homogenization stages in the tungsten-
rhenium system.   Their data also show evidence of the interface turnaround phenomenon discussed previously (Figure 10)
in relation to the two-phase homogenization model [9].   Both
Figures 30 and 31 indicate that the σ:β interface initially
moves away from its center of curvature and subsequently
moves in the direction dictated by the solution of the σ
(and α) phase in the β-phase matrix.

Another method for treating quantitative metallography data for three-phase systems is similar to that discussed previously for two-phase systems:

$$\frac{\xi_\gamma}{\ell} \cong \left(\frac{V_{V\gamma}}{\bar{C}}\right)^{\frac{1}{3}}$$

$$\frac{\xi_\beta}{\ell} \cong \left(\frac{V_{V\gamma} + V_{V\beta}}{\bar{C}}\right)^{\frac{1}{3}}$$

where ξ_γ, ξ_β, $\bar{C}$, and ℓ are defined in Figures 2 and 8. Thus, it is not necessary to carry out average particle radius measurements to obtain the normalized sizes of the β and γ phases and the values of ξ_β/ℓ and ξ_γ/ℓ are directly applicable to evaluation of concentric-sphere models. This approach is currently being applied to the nickel-molybdenum system [12].

X-Ray Compositional Line Broadening. The 311 uncorrected diffraction peaks from a series of nickel-molybdenum samples are shown in Figure 32. A shift in the maximum peak intensity toward $\bar{C} = 0.208$ is noted as well as a decreasing breadth of the peak with increasing time. The intensity near the pure nickel peak for $t = 4.0$ hours and $t = 22.0$ hours is probably due to a non-ideal dispersion of the molybdenum particles which causes islands of relatively pure nickel to remain after long homogenization treatments.

The composition-effective penetration data obtained from the diffraction peaks (after correction) are presented in Figure 33. The corrections were made by using the Fourier method of Stokes [33] and the computer program of De Angelis and Schwartz [34] similar to that used previously [8] for copper-nickel compacts. However, for copper-nickel compacts the whole range of compositions may be analyzed, while for nickel-molybdenum only the nickel-rich solid solution was monitored. The expected shift in the composition-effective penetration data is from a vertical line at:

$$\int_0^C N(C)dC \Big/ \int_0^{0.26} N(C)dC = 1.0$$

(pure nickel) to a horizontal line at $\bar{C} = 0.208$.

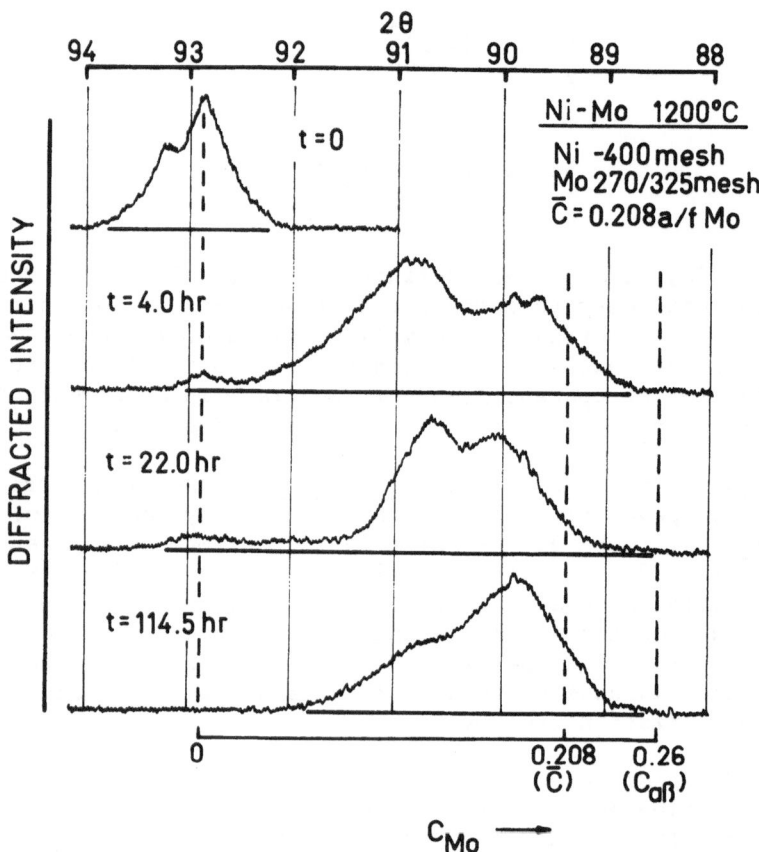

Figure 32. Uncorrected 311 CuK$_\alpha$ diffraction peaks from the
nickel-rich phase in a series of nickel-molybdenum powder
compacts. The pure nickel peak (t = 0) is compared to compacts homogenized for various lengths of time. Broadening
and shifting of the diffraction peak are noted for t = 4.0
hours, followed by a subsequent decrease in broadening but
continued peak shift for increasing time.

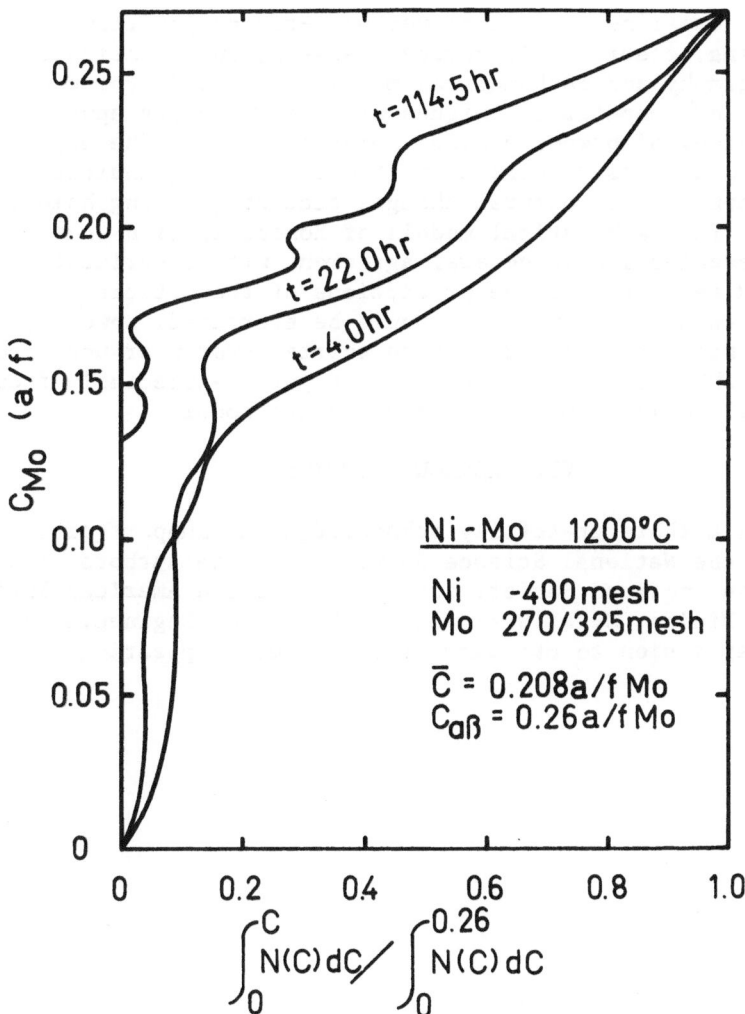

Figure 33. Concentration-effective penetration data obtained from corrected 311 CuK$_\alpha$ diffraction peaks for a series of nickel-molybdenum compacts (Figure 32). The data indicate that, for t = 4.0 hours and t = 22.0 hours, a small amount of pure nickel is present, but for t = 114.5 hours the minimum C$_{Mo}$ is approximately 0.13. The expected trend of the data (to approach a horizontal line at C$_{Mo}$ = 0.208) is observed.

V.   SUMMARY AND CONCLUSIONS

The experimental techniques of x-ray compositional line broadening, electron microprobe analysis, and quantitative metallography and mathematical models developed from numerical methods and computer techniques are complementary approaches to the study of powder compact homogenization. The experimental techniques described provide a direct, quantitative measurement of structural changes occurring during homogenization. The mathematical models of homogenization that have been presented are in general agreement with experimental data and serve to provide predictions of the effects of processing variables. Caution should be exercised, however, in using a mathematical model in an untried system or under conditions where the compact geometry departs appreciably from the ideal geometry used to formulate the model.

VI.   ACKNOWLEDGMENTS

The authors gratefully acknowledge the support of this work by the National Science Foundation. The authors also thank the American Society for Metals and the American Institute of Mining, Metallurgical, and Petroleum Engineers for their permission to use many of the figures appearing in this paper.

REFERENCES

1. B. Fisher and P. S. Rudman, J. App. Phys. 32 (1961) 1604

2. P. Chevenard and X. Waché, Rev. de Met. 41 (1944) 353

3. P. Duwez and C. B. Jordan, Trans. ASM 41 (1949) 194

4. S. Weinbaum, J. App. Phys. 19 (1948) 897

5. S. D. Gertzriken and M. Feingold, Zh. Theor. Fiz. 10 (1940) 574

6. A. I. Raichenko and I. M. Fedorchenko, Dokl. Akad. Naul. Ukr. S. S. R. 3 (1958) 255; 8 (1958) 835

7. A. I. Raichenko, Fiz. metal. metalloved. 11 (1961) 49

8. R. W. Heckel, Trans. ASM 57 (1964) 443

9. R. A. Tanzilli and R. W. Heckel, Trans. Met. Soc. AIME 242 (1968) 2313

10. D. Murray and F. Landis, Trans. ASME (D) 81 (1959) 106

11. R. A. Tanzilli and R. W. Heckel, Trans. Met. Soc. AIME 242 (1968) June

12. R. D. Lanam, Ph.D. Thesis Research, Drexel University, Philadelphia, Pa.

13. J. S. Kirkaldy, Can. J. Phys. 36 (1958) 899, 907, 917

14. G. V. Kidson, J. Nucl. Mat. 3 (1961) 21

15. D. D. Howat, R. L. Craik, and J. P. Cranston, J. Inst. Met. 80 (1951-52) 353

16. J. A. Lund, W. R. Irvine, and V. N. Mackiw, Powder Met. no. 10 (1962) 218

17. J. A. Lund, T. Krantz, and V. N. Mackiw, Progress in Powder Metallurgy, Proc. 16th Ann. Mtg. MPIF 16 (1960) 160

18. F. N. Rhines and R. A. Colton, Trans ASM 30 (1942) 166

19. F. Thummler and W. Thomma, Modern Developments in Powder Metallurgy ed. H. Hausner vol. I Fundamentals and Methods, Consultants Bureau, N. Y. 1966, p. 36.

20. F. Thummler and W. Thomma, Berichte uber de II Internat. Pulvetmet. Tagung (Eisenach 1965) Akademic Verlag, Berlin 1962, p. 137

21. B. Fisher and P. S. Rudman, Acta. Met. <u>10</u> (1962) 37

22. S. B. B. Richardson, A. P. Martin, and B. H. C. Waters, Powder Met. no. 9 (1962) 37

23. K. Torkar and H. Götz, Z. Metallkunde <u>46</u> (1955) 371

24. K. Torkar and H. Neuhold, Z. Metallkunde <u>52</u> (1961) 209

25. M. F. Grimwade, Powder Met. no. 7 (1961) 283

26. U. E. Wolff, Trans. ASM <u>55</u> (1962) 362

27. P. S. Rudman, Acta Cryst. <u>13</u> (1960) 905

28. S. Leber and R. F. Hehemann, Trans. Met. Soc. AIME <u>230</u> (1964) 100

29. R. W. Fraser and D. J. I. Evans, Powder Met. <u>11</u> (1968) 358

30. J. E. White, Trans. ASM <u>57</u> (1964) 756

31. D. W. Smith and R. F. Hehemann, Trans. Met. Soc. AIME <u>336</u> (1966) 506

32. K. F. J. Heinrich, <u>Quantitative Electron Probe Microanalysis</u>, NBS Special Publication no. 298, 1968.

33. A. R. Stokes, Proc. Phys. Soc (London) <u>61</u> (1948) 382

34. R. J. De Angelis and L. H. Schwartz, Acta Cryst. <u>16</u> (1963) 705

35. J. E. Hilliard and J. W. Cahn, Trans, Met. Soc. AIME <u>221</u> (1961) 705

36. R. T. De Hoff, Trans. Met. Soc. AIME <u>224</u> (1962) 474

37. R. A. Tanzilli, Ph.D. Thesis Research, Drexel University, Philadelphia, Pa.

HOT STAGE MICROSCOPY STUDY OF LIQUID PHASE SINTERING

Edul M. Daver and William J. Ullrich

ALCAN Metal Powders*
Division of Alcan Aluminum Corporation
Elizabeth, New Jersey 07207

INTRODUCTION

A substantial part of our knowledge about the structural changes taking place in metals during any form of heat
treatment is obtained by means of the classical method based
on quenching from elevated temperatures followed by metallographic examination. As a result of instrumental and experimental difficulties the direct method of microscopical observation of the specimen during the heat treatment has come
into more general use only during the last two decades. Room
temperature observations can give only a very limited picture of the actual changes taking place. A metallographer
studying a photomicrograph of a cold sample is expected to
guess from circumstantial evidence how the structural changes took place at high temperatures.

This investigation was undertaken to study liquid phase
sintering with the help of the hot stage. The system chosen
for study is 90 copper-10 tin, which is very commonly used
in industry, for example in porous bronze bearings. The instrumentation, technique and problems associated with the
study are discussed in detail. The photomicrographs obtained at elevated temperatures is expected to give more information about liquid phase sintering, in this case of
bronze, than conventional metallographic techniques.

INSTRUMENTATION

The basic instrumentation required in addition to the metallograph is a heating stage; an apparatus to produce and control the electrical power and atmosphere during sintering; and a temperature measuring unit. The metallograph and heating stage used are both Leitz models. A number of different hot stage models exist, though basically all of them operate on the same principle. It is generally constructed of stainless steel and consists of two portions which come together to form a seal. Circulating water in both parts protects the stage from overheating. The sample is heated in a small specimen holder surrounded by a heating element. The specimen holder is generally made of sapphire, sintered alumina, platinum or tantalum and the heating element is usually platinum-rhodium, tungsten or tantalum. Observations and photomicrographs are made through a quartz window in the top of the stage. (Figures 1 and 2.) The features that may vary are, for example, the size of specimen which could be accomodated, maximum temperature which could be utilized and the mechanism used to protect the viewing window.

The temperature producing and controlling device can be bought with the heating stage. When an atmosphere is used, the gas flowing through the unit carries away many BTU's, decreasing the maximum temperature obtainable. Sufficient power must be available to overcome these heating losses and still obtain the desired temperature. It would be useful to keep this in mind as the apparatus is expensive and one would like to operate at the temperature of interest. In this particular case the temperature is controlled by a laboratory set-up which includes a powerstat and a stepdown transformer. To obtain a temperature of $800^{\circ}C$, 20 amps at 24 volts is required. The temperature measuring unit consisted of a thermocouple and galvanometer. A platinum-rhodium thermocouple 0.062" in diameter is inserted through a small opening provided for the purpose, and kept in contact with the side of the specimen. The thermocouple leads are connected to the galvanometer which indicates the specimen temperature. The atmosphere used is tank hydrogen.

One special feature arises from the fact that the specimen surface is at a relatively long distance from the objective and that the two are separated by a quartz window which is itself an optical element. Conventional microscopes use objectives with a high numerical aperture and

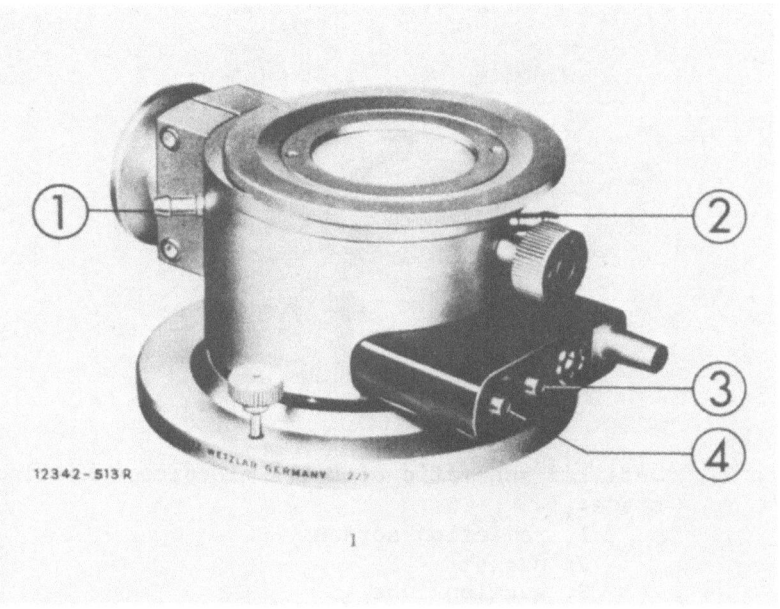

Figure 1.   Leitz microscope heating stage: (1) water out-
             let, (2) and (3) to be bridged, (4) water inlet.

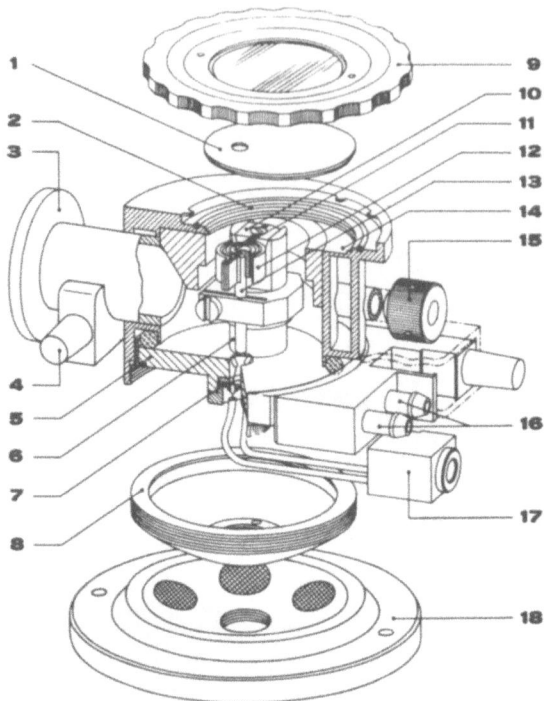

Figure 2.   Detailed schematic of Leitz microscope heating
             stage:
                  1. radiation screen
                  2. gasket
                  3. suction tube
                  4. heating current connection
                  5. base plate
                  6. electrode column with clamp
                  7. thermocouple flange
                  8. screw cup
                  9. cover with quartz window
                 10. clamping wedges
                 11. sample
                 12. heating and screening plate
                 13. capillary tube with thermocouple
                 14. double walled vacuum tank
                 15. flooding valve, inert gas flushing
                 16. cooling water connections
                 17. thermocouple bush
                 18. base with spherical bearings & magnets

correspondingly short working distances to produce high resolution. Those which have sufficient working distances to
be used with a hot stage would have low magnifications with
correspondingly low numerical apertures. Consequently, to
obtain the high resolution at long working distances, a
specially designed objective is used.

An essential part of such a study is the photographic
equipment, accessories and knowledge of color photomicro-
graphy. This will not be discussed here, but those interested could familiarize themselves about it from Kodak's
"Photography through the Microscope," Scientific and Technical Data Book, P-2.

EXPERIMENTAL PROCEDURE

Reduced copper powder (MD 165) and tin powder (MD 101)
are blended together and compacted in a 0.220" diameter die.
No lubricant is used in the investigation to avoid clouding
the viewing window by low temperature volatilization of the
lubricant. The compacted specimen is 0.222" diameter and
about 0.300" high, which corresponds to a green density of
about 6.70 g/cc or 76% theoretical density. One of the flat
surfaces of the cylindrical specimen is polished and etched
lightly with an etchant to reveal the particle boundaries.
The etchant used is a mixture of 50 ml of hydrogen peroxide
and 50 ml ammonium hydroxide. The specimen is then placed
inside the specimen holder and the upper half of the heating
stage is locked into position and the thermocouple placed so
that it touches the specimen. The cooling water is now cir-
culated through the heating stage. Tank hydrogen is passed
through a bubbler at a rate of about 4-5 bubbles/second.
After all the air originally present has been flushed, the
exhaust gas is ignited. The best way to test if the system
has been completely flushed of air is to collect the exhaust
gas in an inverted test tube, and place a lighted match at
its mouth. A pop indicates the presence of air and if the
hydrogen burns, the air has been completely flushed out of
the system. Once the exhaust gas is ignited, the temperature is raised at a rate of approximately 50°C/min. and the
structural changes followed by maintaining the same field of
view throughout the observation. Photomicrographs are taken
when any change in the surface structure is observed. If it
is physically impossible to continue taking meaningful photographs, the hot stage is cooled rapidly and the cause of

intermediate stoppage corrected or the experiment discontinued. Such a situation may arise owing to a number of problems, such as a faulty selection of field being viewed, loss
of field being viewed due to actual movement of specimen,
specimen oxidation or window fogging. All these problems
have to be tackled to obtain good results with minimum loss
of time.

First, the specimen has to be protected against oxidation which would occur rapidly if the specimen were to be
heated in air. The oxide film formed on the metal surface
obscures the structure being observed. Hence, heating is
carried out with the specimen protected either by a vacuum
or by an inert or reducing atmosphere. Sintering of Cu-Sn
mixes generally involves the use of a reducing atmosphere
and for this investigation hydrogen is utilized.

The quartz viewing window must be kept clear at all times
as it is through this window that observations are made. Material liberated from the specimen will condense on the inside surface of the relatively cool window. In addition to
ordinary outgassing, molecules vaporize from the metal surface at elevated temperatures. In vacuum, these molecules
enjoy a long unimpeded travel which facilitates window fogging and further observation is obviously impossible. In
this study, the flow of hydrogen gas seems to sweep the out-
gassed material along with it, thus avoiding window fogging.
No problems are experienced up to about 650°C, above which a
white deposit on the window is occasionally experienced. If
this happens, the hot stage is cooled and the window cleaned.
The deposit was analyzed by means of an Atomic Absorption
unit and found to be zinc. The source of this zinc is the
Cu and Sn powders in which it exists as a trace impurity. The
amount, though very small, seems to be enough to fog the
viewing window. The newer hot stage models have a rotatable
quartz plate much larger than the window opening. When a
section gets clouded, the plate is rotated so that a clean
portion comes over the window opening, simultaneously maintaining a perfect seal.

The method used to reveal structural details is of great
importance. Should the specimen be well polished? Should it
be etched? Should one rely on thermal etching? - or gas etching? A metal surface with a completely mirror-like polish
will act as a reflector and reveal no underlying structure.
Etching techniques used for room-temperature metallography

are useful to a limited extent. Structures obtained by etching at room temperature remain stamped on the polished metal
surface and interfere with recognition of the new structures
which arise during heating and transformation. Thermal etching is generally quite effective at temperatures above 400°C.
Thermal etching works on the principle that at elevated temperatures material at the boundaries is dislodged more rapidly
than within the grain itself. At very high temperatures,
about 750°C and above, the specimen starts to glow appreciably, affecting the contrast which, in turn, controls the
quality of the observed specimen structure. In such cases,
special "gas etching" techniques may prove advantageous. In
the present investigation, conventional light etching is used
at low temperatures which helped considerably to distinguish
particle boundaries and increase color contrast rather than
reveal internal particle structures. At high temperatures,
thermal etching was completely relied on.

The biggest problem with respect to hot stage microscopy of liquid phase sintering is, as the name implies, the
presence of at least one phase being in the liquid state.
The liquid phase accelerates sintering and structural changes
take place rapidly. It is quite normal for the resulting
sintered piece to have a rough surface and undergo considerable changes in its dimensions. Hence, the area under observation has to be continuously watched, tracked and refocused.
As soon as one of the phases melts, it loses polish completely and a negligible amount of light can be reflected from the
liquid surface, thus appearing black under the microscope.
As the temperature is raised, the liquid spreads in all directions and if a tin rich area has been chosen for observation the whole field of view will appear black. The liquid
phase, however, reacts simultaneously with the solid phase
forming intermediate phases. These intermediate phases,
though solid, will reflect far less light than a polished
specimen and, once again, appear relatively dark. The intermediate phases solidify at slightly different levels resulting in out-of-focus regions.

The heating is continued, and photomicrographs taken at
regular intervals, up to a temperature of 800°C where it is
held for 10 minutes. The photomicrographs thus taken are
those in Figure 3. The series enables us to actually see
the sintering process take place. We are able to follow the
manner in which the tin melts, spreads, reacts with the copper particles, forms intermediate phases ending with a solid

Figure 3.　Photomicrographs, using direct heat-up technique, of 90 Cu-10 Sn compact, green density of 6.7 g/cc; polished and etched lightly with H_2O_2 and NH_4OH; field of view approximately 375 x 280 microns. (All 250 x) (Reduced 10% for reproduction.)

Figure 4.　Photomicrographs, using intermediate examination technique, of 90 Cu-10 Sn compact, green density of 6.7 g/cc; field of view approximately 375 x 280 microns.　(All 250 x) (Reduced 10% for reproduction.)

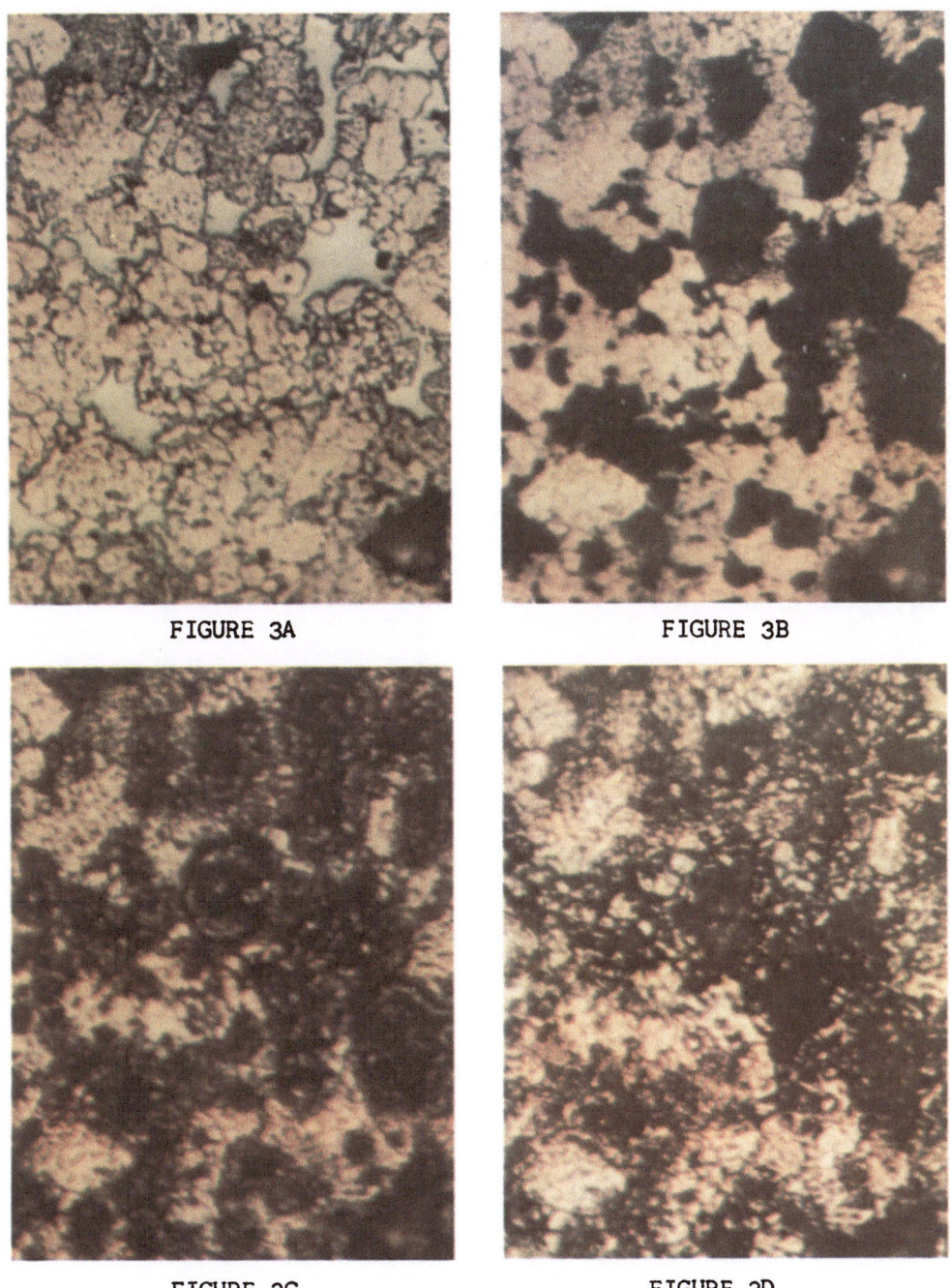

FIGURE 3A

FIGURE 3B

FIGURE 3C

FIGURE 3D

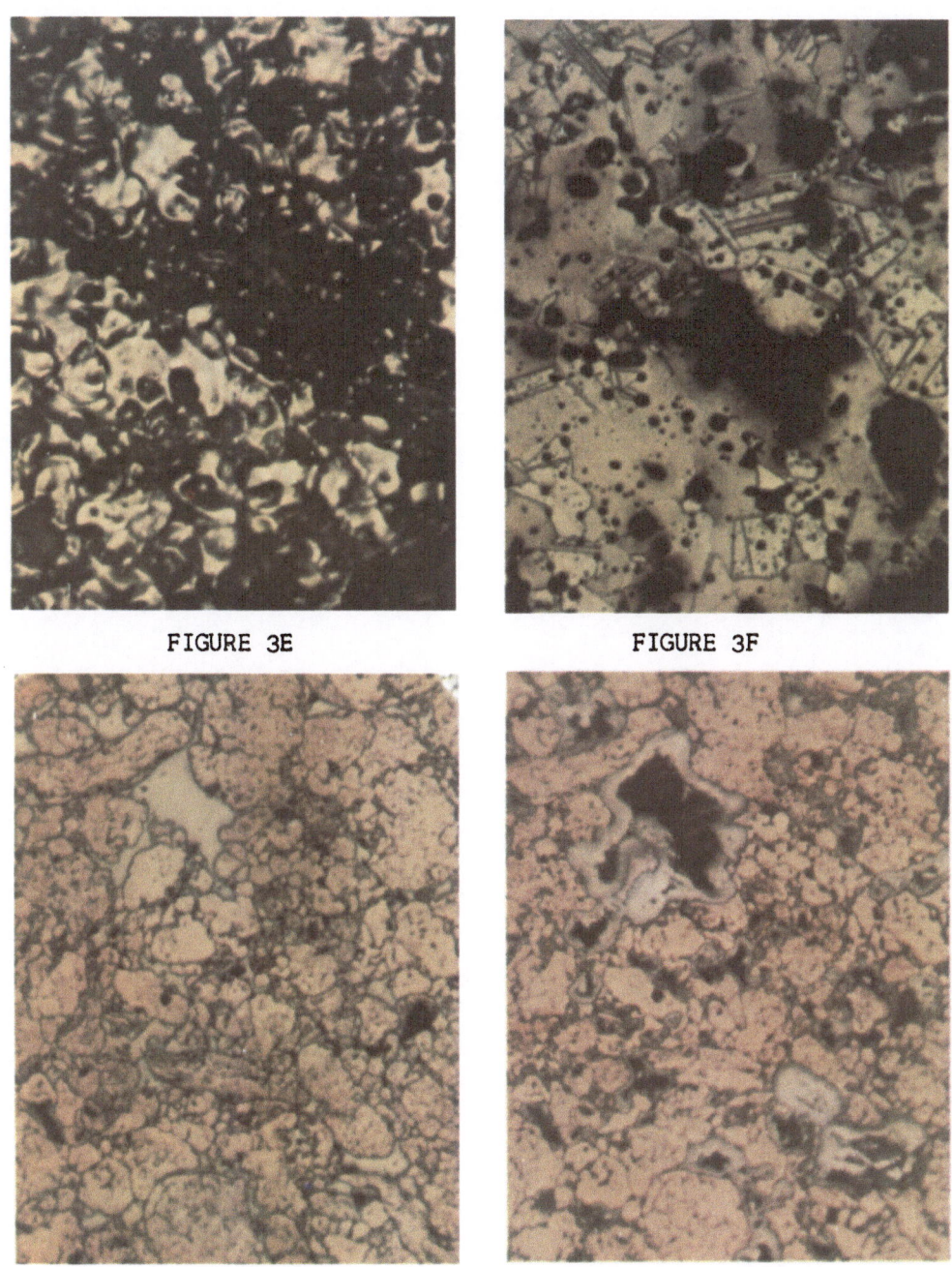

FIGURE 3E   FIGURE 3F

FIGURE 4A   FIGURE 4B

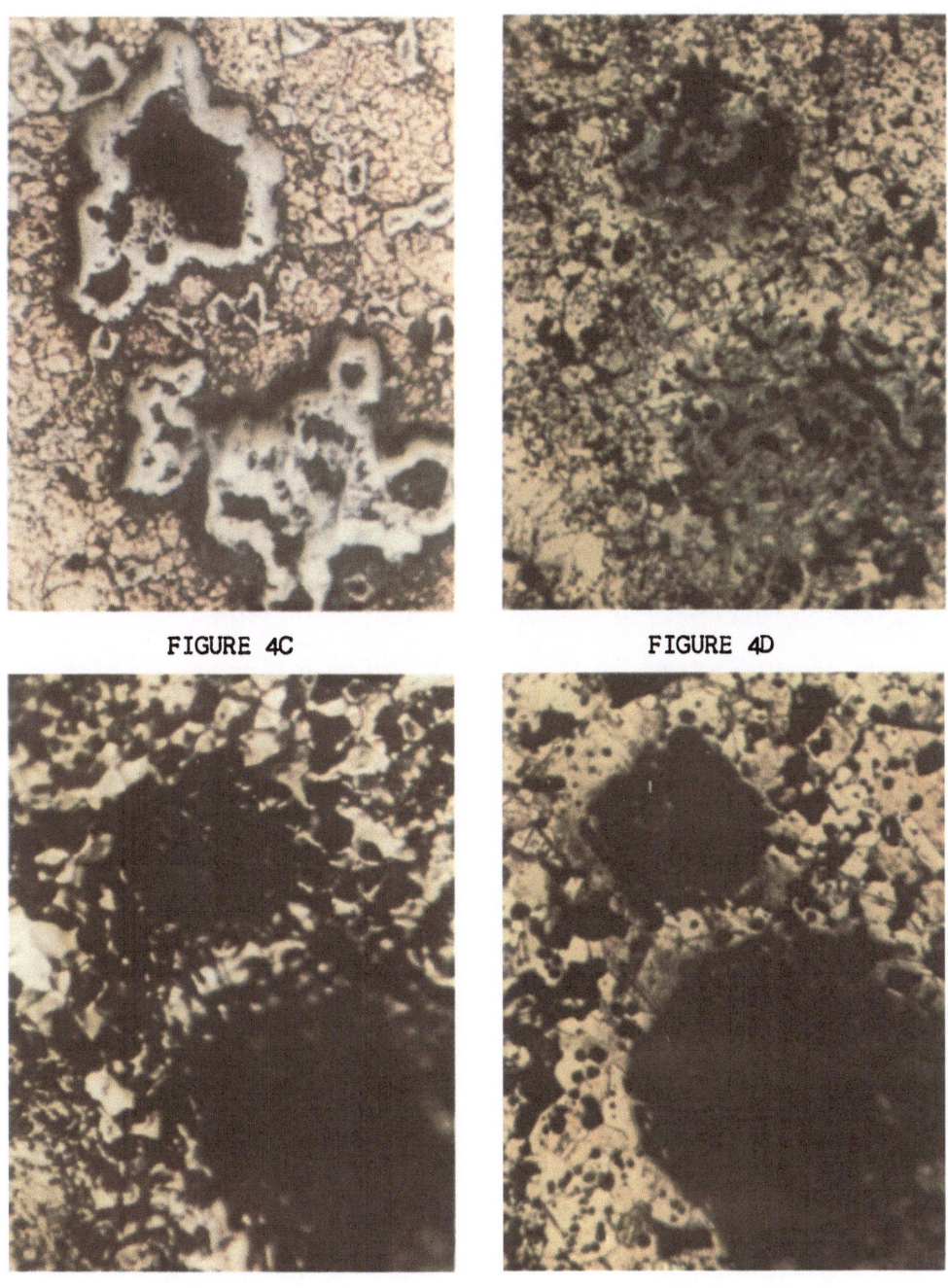

FIGURE 4C

FIGURE 4D

FIGURE 4E

FIGURE 4F

solution of alpha-bronze. It does not, however, enable us
to observe in detail where and how the intermediate phases
form; the way in which crystals form; the creation of porosity and its change in shape. Most important of all, it
does not allow identification of intermediate phases. In
short, it allows us to see macroscopically the process of
liquid phase sintering though not allowing a microscopical
study.

The procedure was varied slightly to obtain more structural details. As soon as the tin melts and spreads, the
specimen is cooled, polished and etched lightly and then re-
heated with the same field of view being maintained. This
procedure was repeated at intervals of I00°C. Using the hot
stage, the same field of view can be maintained and the temperature at which changes are taking place accurately regis-
tered. With this known temperature and the Cu-Sn phase diagram, the phase existing is identified. The photomicrographs
taken using this procedure are those in Figure 4. This procedure gives us far more detailed information than the direct
heat-up though it does not represent an actual practical case.

The investigation using both the methods described gives
us first hand knowledge of what actually happens during liquid phase sintering.

RESULTS & DISCUSSION

Figure 3: Direct Heat-Up Technique

In Figure 3a the light brown particles are copper, the
grey particles are tin, and the black regions indicate porosity. This is the room temperature structure.

In Figure 3b the temperature is 400°C. The tin particles have melted and the liquid started to spread. Under the
microscope, the liquid tin areas appear black owing to loss
of reflectivity. The pores and tin areas are indistinguish-
able except by comparison with the previous photomicrograph.
Liquid tin reacts with the copper particles forming eta and
epsilon intermediate phases. The intermediate phases are,
however, not identifiable in the photomicrograph.

In Figure 3c the temperature is 600°C. The amount of

liquid phase has decreased considerably, resulting in increased intermediate phases which are clearly visible now.
An appreciable number of copper particles are as yet unreact-
ed. Thermal etching is evident at this temperature.

In Figure 3d the temperature is 800°C. The temperature
is above the peritectic temperature of 798°C. Whatever beta
is present forms a liquid phase above this temperature. The
copper phase is now separated from the liquid only by the
alpha phase and a remarkable promotion of diffusion is evident. The yellowish alpha grains, which lie between the dark
liquid areas and the brown copper particles, are clearly visible in the photomicrograph.

In Figure 3e the temperature is held at 800°C for 10
minutes. Concentration leveling, homogenization and solidification has taken place rapidly. Grain growth and pore
spherodization is also evident.

In Figure 3f the specimen is cooled, polished lightly
and etched. A uniform microstructure consisting of alpha
grains, a number of them twinned, is obtained.

Figure 4: Intermediate Examination Technique

In Figure 4a the 90 Cu-10 Sn green compact is at room
temperature. The light brown particles are copper, the grey
particles are tin, and the black regions indicate porosity.

In Figure 4b the temperature is 300°C. The specimen is
cooled, polished and etched so that the structural changes
could be followed and intermediate phases recognized. As
soon as the tin melts, it reacts with the copper particles
forming intermediate phases - first eta (60% Sn) followed by
epsilon (38% Sn). The photomicrograph indicates clearly
that the intermediate phase is formed at the interface and
not in the tin melt. The areas originally occupied by the
tin phase have left behind pores of approximately the same
size and shape.

In Figure 4c the temperature is 500°C. The specimen is
cooled, polished and etched. The liquid phase has spread rapidly. Notice the appearance of a second large intermediate
phase area, which indicates the presence of a tin rich area
just below the surface being observed. The unattached copper

particles are clearly distinguishable from the intermediate
phase. The alpha and delta phases are not recognizable as
yet.

In Figure 4d the temperature is 700°C. The specimen
is cooled, polished and etched. Crystal formation and metallurgical bonds developed rapidly. The phases now existant
are alpha and delta. The delta phase (blue) is retained at
room temperature as it decomposes to alpha plus epsilon only
under complete equilibrium conditions.

In Figure 4e the temperature is 800°C. The temperature
is above the peritectic temperature. Diffusion increases
tremendously and is accompanied with concentration leveling,
homogenization and solidification.

In Figure 4f the temperature is 800°C. The specimen is
cooled, polished and etched. The structure is alpha bronze
with evidence of twinning, pore migration and spherodization.
The two large pores are as a result of, what is called in
industry, "tin errosion" and is commonly experienced during
sintering of bronze.

SUMMARY

Sintering processes are generally studied with the help
of conventional metallography. The investigator, thus, makes
his conclusions from room temperature structures and can
never be sure of what actually happens during sintering. Hot
stage microscopy is being used, to a limited extent, for the
study of solid phase sintering. The present study shows that
hot stage microscopy can also be used successfully to study
liquid phase sintering. The technique is useful in observing the manner in which the lower melting point constituents
melt, spread, and react with the other constituents forming
intermediate phases and finally a homogeneous alloy. It is
also useful in following the formation of porosity and/or
its change in shape as sintering proceeds. The copper-tin
system which is being studied will hopefully allow us to
understand and eliminate certain unsolved problems experienced; for example, in the manufacture of bronze bearings. The
major advantage of hot stage microscopy as compared with conventional metallography is that the specimen is actually
viewed at the elevated temperatures and thus the structural
changes are observed "live." Photography is used to provide

a continuous record of the transition process. It is therefore certain that in the future the technique of hot stage
microscopy as applied to liquid phase sintering will enable
us to make definite statements regarding a number of specu-
lative and inconclusive existing theories.

DYNAMIC ELASTIC DETERMINATION OF THE PROPERTIES OF SINTERED POWDER METALS

R. H. Brockelman

Army Materials and Mechanics Research Center

Watertown, Massachusetts   02172

INTRODUCTION

Powder metallurgy products normally contain an inherent amount of residual porosity which gives the product reduced density. Sintered density is a commonly specified characteristic of powder metallurgy products, because, to a first approximation, mechanical properties are related to density. Often, however, this is a poor approximation due to the recognized fact that porosity, per se, does not determine the properties of the product; that is, at any level of density, properties may vary over a relatively wide range. The size, shape and distribution of the porosity, in addition to interparticle bonding also contribute to the mechanical properties[1,2,3]. These factors are controlled by a multiplicity of processing variables that are difficult for the user to specify or monitor.

The inherent potential for variability obliges the user to adopt stringent acceptance test procedures in order to insure product serviceability. For this purpose, actual or simulated service-testing schedules are, and will continue to be, indispensable. However, in cases where high reliability must be maintained, the application of valid auxiliary nondestructive tests can be of greater value. A program, designed to ultimately establish such tests, is the subject of this paper.

If the composition and thermal treatment of a powder

metal product are held constant, the mechanical properties
of the product should be related to the disposition of
residual porosity[4]. The problem is to select a method of
interrogating the material that is sensitive to the nature
of this residual porosity, yet affords some suppression of
irrelevant variables. Two basic dynamic methods, mechanical
resonance and ultrasonics, appear to meet these requirements.

The concept of dynamic testing is not new. It origi-
nated when the "soundness" of material was crudely determined
by tapping the part and listening to the ring. The possible
application of these methods to quality control of powder
metal products was pointed out as early as 1950[4] but has
apparently received little attention. Dynamic elastic
measurements have been applied, however, to a number of other
porous or mechanically heterogeneous materials. Sonic
testing has been recognized for several years as a means for
studying the quality of concrete[5]. Resonant frequency
measurements have been used to control the quality of abrasive
wheels[6] and have been applied to the nondestructive evaluation
of gray and nodular cast irons[7]. Ultrasonic velocity techniques have been used to evaluate the mechanical properties
of bulk graphite[8] and to measure the moduli of sintered polycrystalline ceramic compounds such as MgO and Al_2O_3.[9] These
applications are of interest because the structures of these
materials are quite analogous to those found in powder metal
products.

ELASTIC MODULI AND VELOCITY OF SOUND PROPAGATION

The use of dynamic elastic techniques as potential
test methods for sintered powder metals is based on the
theoretical relationships between the velocity of sound
propagation in a material and its elastic modulus.

Equations relating the natural resonant frequency and
Young's modulus of uniform cylindrical and rectangular bars
of homogeneous, elastically isotropic materials have been
presented by Spinner and Tefft[10]. In generalized form, the
resonant frequency of a body has been written as[6]

Frequency = shape factor X physical constants factor   (1)

For cylinders and rectangular specimens having a length
much greater than the cross section, the fundamental longi-

tudinal resonant frequency is

$$f_1^L = \frac{K}{2l} \sqrt{\frac{E}{\rho}} \tag{2}$$

Where

f_1^L = fundamental longitudinal resonant frequency in cps.

l = length of the sample in inches.

E = dynamic Young's modulus in psi.

ρ = density in g/cc.

K = a constant the value of which depends on the units. For the units

$$K = 103.4 \ \frac{g^{1/2} \cdot in^2}{lb^{1/2} \cdot cm^{3/2} \cdot sec}$$

For very accurate measurements or when dealing with relatively short samples, a correction factor involving Poisson's ratio may also be used[10].

The wavelength (λ) of the fundamental longitudinal resonant frequency is related to the sample length by

$$\lambda = 2l \tag{3}$$

from which the thin-rod sonic velocity may be calculated.

$$V_O = f_1^L \lambda = 2f_1^L l = K \sqrt{\frac{E}{\rho}} \tag{4}$$

Where V_O = the thin-rod sonic velocity in in./sec.
Sound velocities may also be measured by ultrasonic techniques. In this case the "bulk" velocity (V_L) is measured which is given in in/sec by

$$V_L = K \sqrt{\frac{E}{\rho}} \left[\frac{(1 - \sigma)}{(1 + \sigma)(1 - 2\sigma)} \right] \tag{5}$$

Where σ = Poisson's ratio.
The ratio of the ultrasonic bulk velocity to the resonant
thin-rod velocity is

$$\frac{V_L}{V_0} = \sqrt{\frac{(1-\sigma)}{(1+\sigma)(1-2\sigma)}} = f(\sigma) \qquad (6)$$

From the foregoing it is seen that both the thin-rod and
ultrasonic velocities are primarily dependent on $\frac{E}{\sigma}$ because
the values of f (σ) for sintered iron range from 1.10 for a
Poisson's ratio of 0.25 to unity as the Poisson's ratio
approaches unity.

POROSITY AND MODULUS

The effect of porosity on the modulus of an otherwise
homogeneous material has been the subject of a number of
theoretical and experimental investigations. An extensive
bibliography on this subject is available and it shows that
the effective modulus of a porous material is related to the
shape and distribution of porosity in addition to the total
void fraction.[11,12,13,14]

The established dependence of Young's modulus on the
nature of the porosity, which also determines the mechanical
properties of a sintered powder metal product, leads to the
hypothesis that sound velocity measurements should be indicative of mechanical properties. In fact, for those cases
where porosity is the only variable, the determination of
modulus may constitute an indirect nondestructive test for
strength.

METHODS OF DETERMINING MECHANICAL RESONANT FREQUENCIES

The techniques and equations that will be described are
applicable to any material that is homogeneous, of the
appropriate shape, and sufficiently elastic to be vibrated.
If the material contains porosity and that porosity is
randomly distributed, the condition of homogeneity is fulfilled macroscopically. If the conditions of proper shape
are not fulfilled, then the various resonant frequencies may
be determined and compared for quality control even though
elastic moduli are not calculated from them.

The sample to be measured is supported on adjustable cross wires and is driven by electromagnetic or piezoelectric transducers. Two systems are available with this instrument which is called the Elastomat for the excitation of the test specimens to its resonant frequencies. It may be vibrated by means of a variable frequency oscillator which is driven either manually or by a motor. The transducer converts the electrical oscillations into mechanical vibrations and couples them to the sample. At the other end, the mechanical vibrations of the sample are reconverted into electrical signals by the receiving transducer. These electrical signals are amplified and their amplitude is indicated on the meter and displayed on the oscilloscope screen. At the resonant frequency, the amplitude of vibration increases to a maximum value, and this frequency is measured by a high-precision counter. However, it is not always necessary to search for the resonant frequency by varying the frequency oscillator. Another system is available in which self-excitation of the sample by means of a positive feedback system is used. The test sample is the frequency-determining element in this arrangement. This method has an advantage in long term experiments, such as modulus versus temperature measurements, because the frequency does not require re-adjusting. In fact Marlowe and Wilder[15] have studied sintering kinetics by monitoring changes in resonant frequency of a bar during sintering.

Using this equipment it is possible to resonate a sample in different ways - longitudinally, flexurally and torsionally. Young's modulus is related to the longitudinal and flexural resonant frequencies and the shear modulus is related to the torsional.

Figure 1 shows electromagnetic transducers arranged so as to produce longitudinal sonic vibrations in a test bar. Notice the electromagnetic transducers do not contact the bar but are positioned approximately 1/4 inch away from its ends. The vibration is created in the bar by means of a varying electromagnetic force. When the excitation and resonant frequency coincide, the bar will oscillate. This transducer is restricted, of course, to ferromagnetic materials. In order to excide nonmagnetic materials with electromagnetic transducers it is necessary to either cement a suitable magnetic disc or apply a magnetic paint to the ends of the sample. In the longitudinal mode the sample is subjected to an alternating tensile compressive force along its length axis as shown in this Figure. The equation relating the fundamental

$$f_{\ell}^{L} = \frac{1}{2\ell} \sqrt{\frac{E}{\rho}}$$

Figure 1. Longitudinal resonance: experimental arrangement (top); mode of vibration (middle); general equation (bottom).

longitudinal resonant frequency and Young's modulus is also
shown. The general equations of Figures 1, 2, and 3 are
presented only for the purpose of demonstrating the relationships of the various resonant frequencies and their elastic
moduli. For the basic equations, required for computing the
elastic moduli from resonant frequencies, see Reference 10.

Arrangement of the transducers over the ends of the bar
as shown in Figure 2 will place the bar into its flexural
fundamental resonant frequency. In this case, the ends of
the specimen are oscillating in the same direction at the
same time, while the center is oscillating in the opposite
direction. Strain amplitude is greatly exaggerated in these
sketches. The maximum strain amplitudes for these methods
are usually 10^{-5} or 10^{-6}. The flexural resonant frequency
is also a function of Young's modulus. In the equation
shown, a = the cross-sectional dimension in the direction of
vibration. The numerical value of the fundamental flexural
frequency is approximately 1/6 of the longitudinal resonant
frequency.

Figure 3 illustrates the method of measuring the torsional resonant frequency, which is a function of the Shear
modulus (G) or Modulus of Rigidity. It is necessary to use
the piezoelectric method to excite the torsional resonant
frequency. Piezoelectric crystals convertsonic energy into
electrical energy and vice-versa. To them are attached very
fine stainless steel wires which in turn are affixed to the
specimen by a small piece of wax or glue. With the wires
perpendicular to the axis of the specimen, it is possible to
easily produce torsional vibrations. The piezoelectric method
can also be used to excite longitudinal and transverse oscillations.   In the fundamental mode of torsional resonance the opposite ends twist in opposite directions at
the same time. From the equation it is seen that frequency
is a function of the shear modulus instead of Young's modulus.

The sizes and shapes of the parts that can be resonated
by such techniques usually depend on the material being tested.
The dimensions of the parts should be large enough so that
their different fundamental resonant frequencies will be
within the range of the equipment. The Elastomat can be used
to test specimens with resonant frequencies up to about
50,000 cycles/sec. For rectangular bars the ratio of length
to either cross-sectional dimension should probably not be
less than 3 to 1. Cylindrical rods with length to diameter

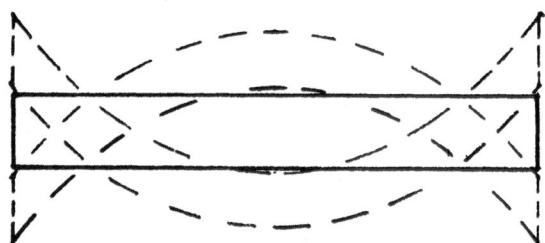

$$ f_l^F = \frac{a}{l^2} \sqrt{\frac{E}{\rho}} $$

Figure 2. Flexural resonance: experimental arrangement (top); mode of vibration (middle); general equation (bottom).

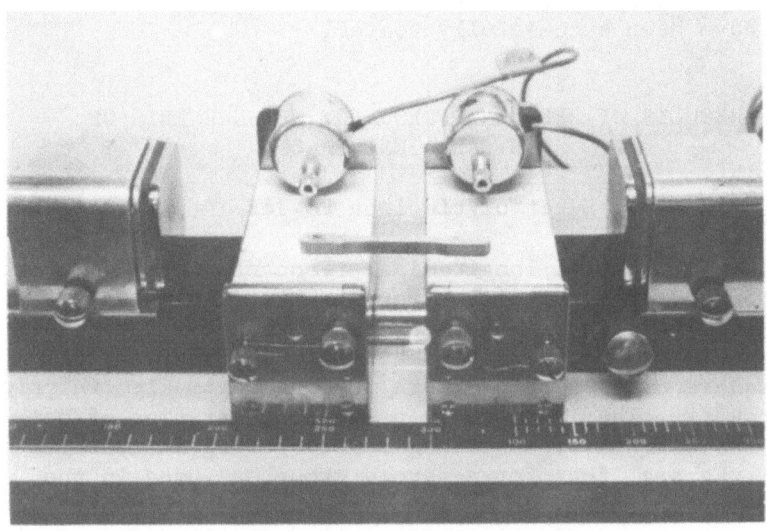

$$f = \frac{1}{2\ell} \sqrt{\frac{G}{\rho}}$$

Figure 3. Torsional resonance: experimental arrangement (top); mode of vibration (middle); gneral equation (bottom).

ratios of 2 to 1 have been tested. Ring shaped sintered
powder metal parts have also been resonated. One such part
was the sintered iron rotating band for the 105 mm artillery
shell. Other specimens with densities as low as 70% theoretical have been successfully tested.

RESULTS OF RESONANT FREQUENCY MEASUREMENTS OF SINTERED TEST BARS

Electrolytic Iron Powder

The fundamental longitudinal resonant frequencies of
two groups of test specimens were measured using the
commercially available equipment described earlier. The
test specimens, approximately 2 1/2 inches long and 1/2
inch square cross section, were prepared from electrolytic
iron powder. The iron powder used in preparation of one
group was cleaned by annealing for two hours in hydrogen at
500°C, whereas, that for a second group was used in the "as
received" condition. Each group contained specimens pressed
in the transverse direction to green densities of approximately 5.5, 6.0, 6.5, and 7.0 g/cc and sintered at 1100°C or
1250°C for periods of 1/2 or 2 hours.

The processing parameters were so designed that specimens of similar sintered density would be produced by different combinations of compaction and sintering. The sintered
densities of the test specimens were determined by dry volume
measurements.

The experimental values of fundamental resonant frequency were used to obtain the thin-rod sonic velocity in
accordance with equation (4). Velocity comparisons which
require a knowledge of sample length and resonant frequency
were used because the length of the samples was not constant.
If, however, components of the same dimensions are to be
evaluated, the resonant frequency may be compared directly.

Figure 4 shows the tensile strength of the specimens as
a function of their sintered density. Tensile strength
generally increases with density for compacts subjected to
the same sintering treatment, and with sintering temperature
and time for specimens of equivalent density. Hydrogen
annealing also produces a moderate gain in tensile strength.
The overall result is a general increase in tensile strength

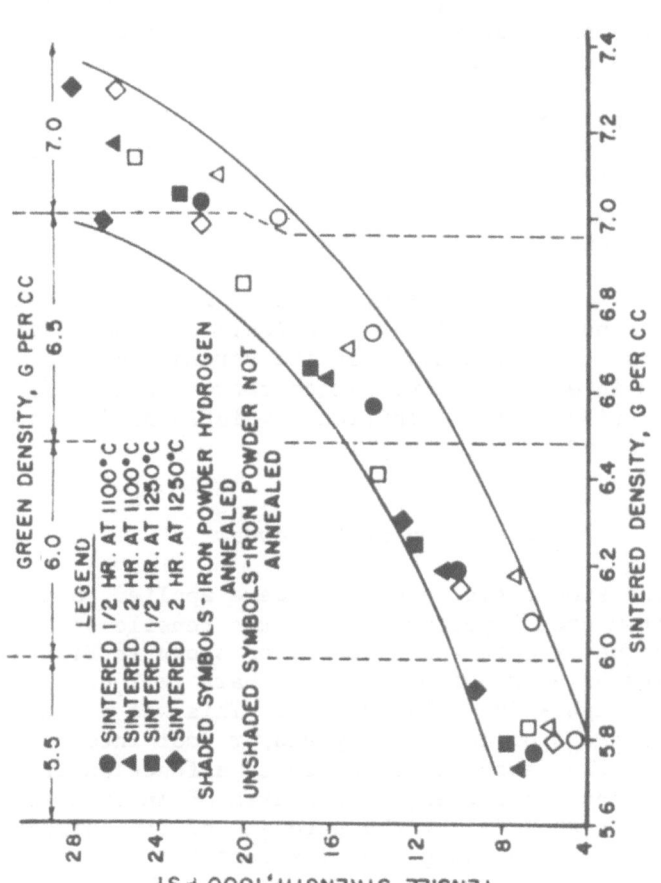

Figure 4.   Dependence of tensile strength on sintered density.

with increasing density but, due to the processing variations, a wide range of tensile strength at equivalent density is obtained. For example, the tensile strength varies from 18,000 to 27,000 psi at 7.0 g/cc sintered density. This demonstrates the inadequacy of sintered density as a criterion of powder metal quality.

If the tensile strength is plotted against the thin-rod sonic velocity of these same samples, a much better correlation is obtained as shown in Figure 5.

A semi-log plot of the same data (tensile strength on the log scale) (Figure 6) results in a straight-line relationship of the following form:

$$\text{Log T.S.} = 1.07 \times 10^{-5} \, V_0 + 2.37 \qquad (7)$$

A comparison of Figures 4, 5, and 6 shows that the velocity of sound propagation, as calculated from resonant frequency, is a better measure of the strength of the material than is the sintered density. Furthermore, it should be noted that the relation is not significantly influenced by variations in processing.

Prealloyed Steel Powder

When the same sonic techniques were applied to steel bars made from prealloyed powder and the tensile strength-resonant frequency data were plotted as shown in Figure 7, the strengths of a number of the test bars fell conspicuously below the straight-line relationship. This was attributed to localized areas of low density and/or poor interparticle bonds. Since the resonant frequency is a function of the entire component properties, a weak area of an otherwise uniform compact could alter the strength of the compact considerably and, depending upon its location and size, not appreciably affect its resonant frequency. These nonuniformities, which probably originate during the pressing process, were not evident in the investigation on sintered iron compacts. The result is anticipated when the relative ease of compacting iron powder compared to prealloyed steel powder is considered. The importance of a method capable of detecting nonuniformities   in   sintered compacts is thus realized.

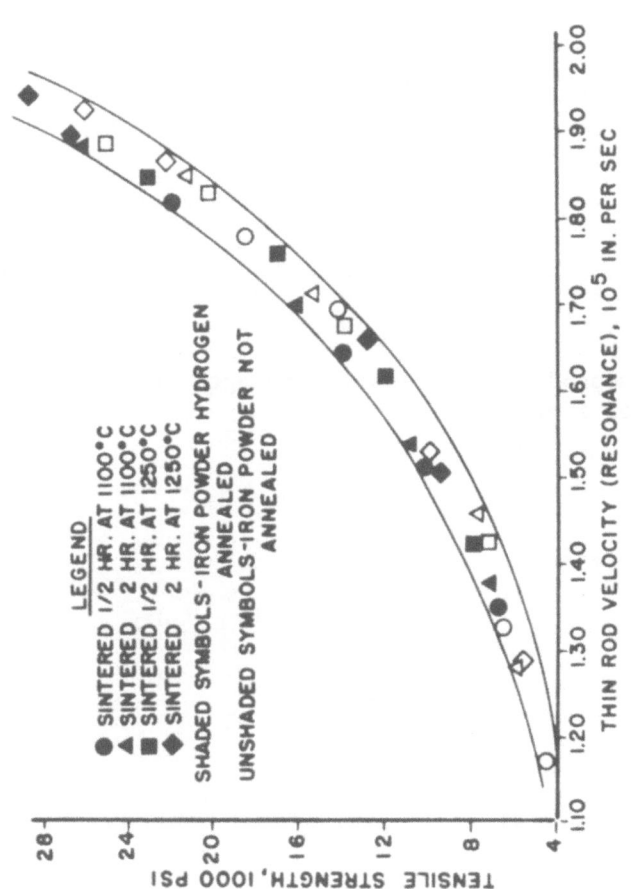

Figure 5. Dependence of tensile strength on thin rod velocity.

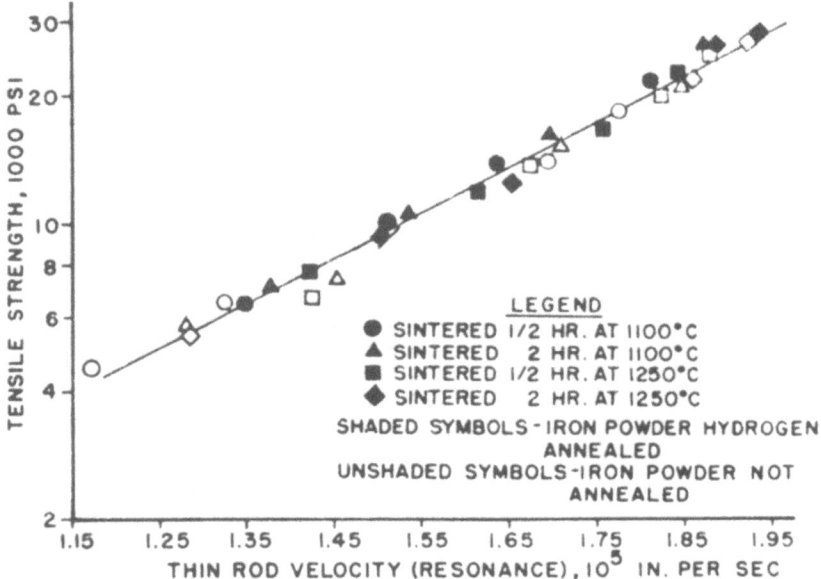

Figure 6. Dependence of tensile strength on thin rod velocity for iron.

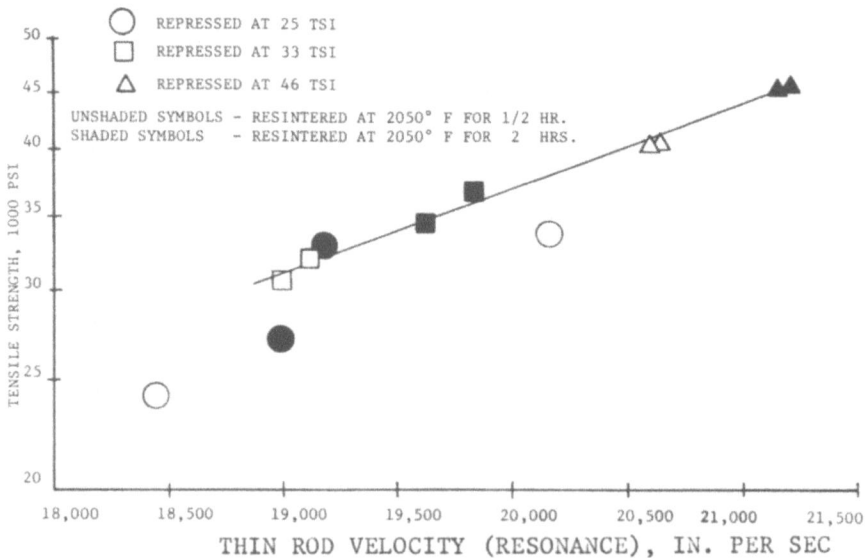

Figure 7. Dependence of tensile strength on thin rod velocity for 4650 steel.

PULSED ECHO TECHNIQUE OF
ULTRASONIC VELOCITY MEASUREMENT

Ultrasonic longitudinal bulk velocity measurements were
also made on the sintered iron test specimens by the pulse-
echo method using a 1/4 inch 2.25 megacycle crystal transducer coupled to the specimen surface with a layer of oil.
It is desirable to seal the surface of the specimen in the
area where the couplant is to be applied. For this test a
clear acrylic lacquer was sprayed on the surface to prevent
infiltration of the liquid couplant.

In the pulse-echo method of testing an extremely short-
time pulse of high-frequency electrical energy is produced
by a pulse generator and applied many times per second to a
piezoelectric transducer. The transducer converts the
electrical pulses into mechanical vibrations which are
transmitted to the specimen through the coupling medium. In
a specimen with parallel surfaces placed perpendicular to the
ultrasonic beam, the ultrasonic energy is reflected back and
forth until the total impinging energy is dissipated. Each
return to the transducer is reconverted to an electrical
signal, amplified and displayed on a cathode ray tube as a
vertical pip or echo. If no discontinuities exist between
the test surface and the opposite face, only the initial
pulse and a series of back echoes will be displayed on the
CRT.

The pulse that appears on the CRT display can be either
the high frequency oscillations that make up the pulse or
these may be rectified for a simpler presentation so that
only the upper half of the pulse shape is displayed as a
pulse envelope. The latter type, often referred to as a
video presentation, was used in this study for velocity
measurement. The pulses are sent out many times per second
so that sufficient image intensity is built up on the cathode
ray screen by repetition. The rate at which they are sent
out is synchronized with the sweep of the CRT; that is, when
a pulse is emitted, the sweep starts. Echoes appear on the
screen at distances proportional to the travel times between
the reflecting surfaces.

One of the simplest methods of velocity determination
uses a continuously variable oscilloscope sweep delay to
measure the time interval between peaks of successive, detected and rectified echoes. A block diagram of the basic

equipment required for this method of velocity measurement
is shown in Figure 8.   In this study an Automation Industries
UM700 Reflectoscope was used as both the pulser and receiver
and a Tektronix 535A oscilloscope was used for the time
interval measurement.   The start of the horizontal sweep of
oscilloscopes equipped with the delayed sweep feature, such
as the 535A, can be continuously delayed by the Delay Time
switch and the Delay Time Multiplier control.   The settings
of the two controls are multiplied together to obtain the
actual delay time.   To measure the time interval between
two echoes the Delay Time Multiplier dial is adjusted to
horizontally position the center of the peak of the first
echo to one of the vertical lines on the scope graticule.
For precise measurement the Delay Time switch should be set
at the fastest sweep rate.   After recording the reading of
the Delay Time Multiplier dial the second echo is positioned
to the same vertical line and the dial position is again recorded.   The time interval between echoes is then calculated
by subtracting the first reading from the second and multiplying by the setting of the Delay Time Switch.   Several
other pulse-echo methods of ultrasonic velocity measurement
are described by Truel, Elbaum and Chick.[16]

To avoid zero errors, it is recommended that the distance between transmitting pulse or initial pulse and first
echo not be measured, and, instead one or several distances
between successive echoes.   This time interval represents
the time required for the pulse to travel twice through the
specimen thickness.   Since the thickness of a specimen can
be measured, the bulk velocity (V_L) is then easily calculated from

$$V_L = \frac{2T}{t}$$ (8)

Where   T = specimen thickness

t = the measured interval between reflected pulses

This method does not result in absolute sound velocity
values of high accuracy.   It does, however, satisfactorily
detect differences in sound velocity in sintered materials.

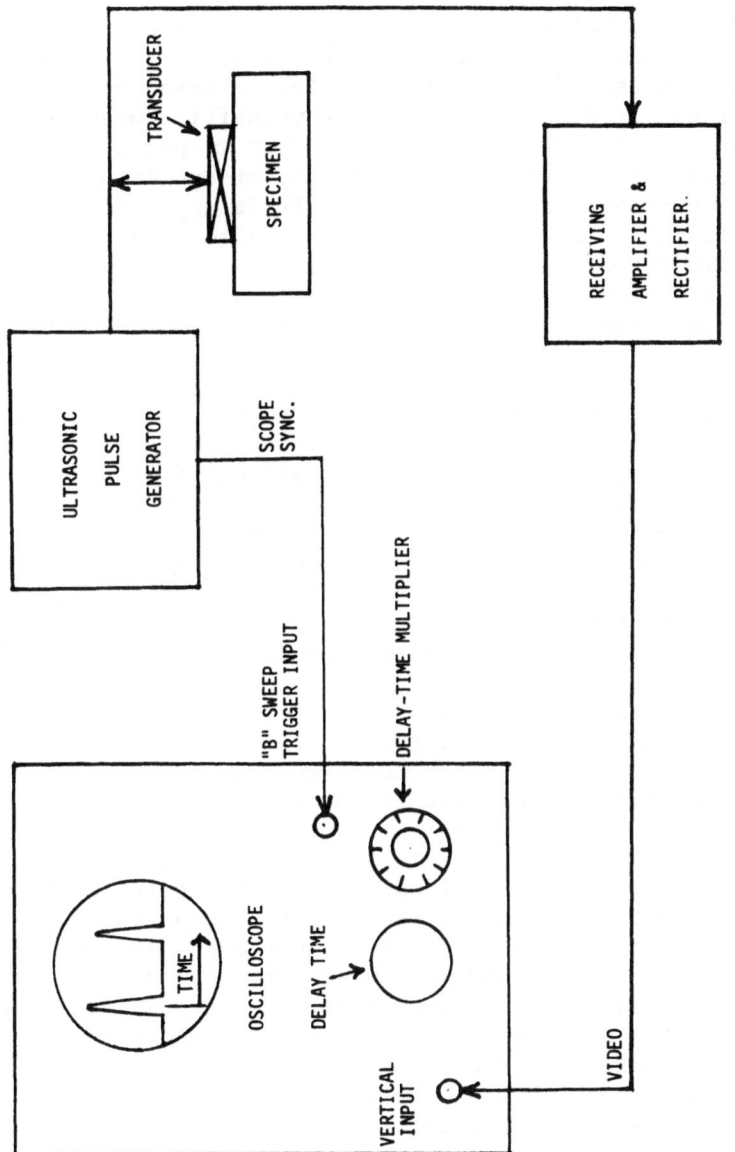

Figure 8. Block diagram of equipment for measurement of ultrasonic velocity by pulsed echo technique.

RESULTS OF ULTRASONIC VELOCITY MEASUREMENTS
OF SINTERED IRON

Figure 9 illustrates the type of correlation that is obtained when tensile strength and ultrasonic velocity are plotted in the same form as Figure 6 for thin-rod velocity. The data shown in Figure 9 represents velocities measured in each of the transverse directions, that is parallel and perpendicular to the pressing direction for specimens prepared from hydrogen-annealed powder. It is apparent that the longitudinal ultrasonic velocity can also be correlated with tensile strength. However, because velocity in the pressing direction for this material is lower than in the perpendicular direction, two lines are shown in the plot. The reason for the directional dependence of velocity is apparent when the microstructures are considered. The porosity in green compacts was aligned nearly perpendicular to the pressing direction, a feature which persisted until the latter stages of sintering. This alignment of pores produced a reduction in cross-section in the pressing direction with a corresponding decrease in modulus, which accounts for the fact that the ultrasonic velocity was lowest in the pressing direction. More complete sintering caused the pores to become nearly spherical in shape, and the difference in ultrasonic velocity with direction diminished. The two lines in the plot intersect at the high strength levels showing the velocity and hence the modulus is nearly the same in both transverse directions. The necessity of making all ultrasonic velocity measurements in the same relative direction is realized if correlation with tensile strength is to be achieved.

Although the cross-sectional area of the pores or inter-particle bonds usually varies relative to the pressing direction, the direction of greatest pore area has been found to be largely dependent on the powder particle shape. For example, in sintered, prealloyed steel powder compacts, where the steel powder appeared nearly spherical, the interparticle bonding was greatest in the pressing direction and ultrasonic velocities were high in the compacting direction for this material. The specific character of each material must be considered independently when analyzing the results of the ultrasonic velocity measurements. It is obvious that ultra-sonic velocity measurements can be used to evaluate isotropy of the material or to infer pore shape and orientation, as well as determine strength.

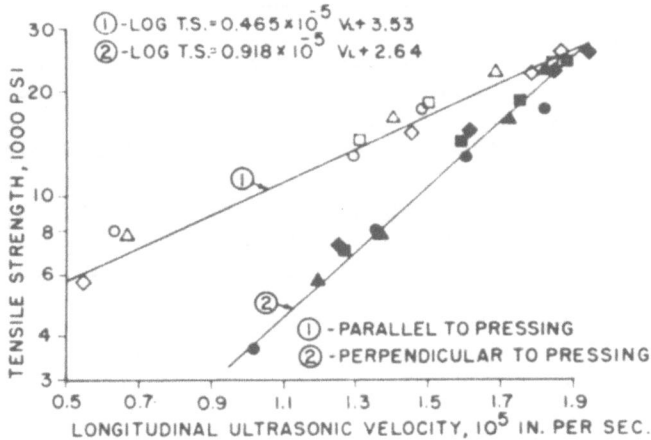

Figure 9.   Dependence of tensile strength on longitudinal ultrasonic velocity for two directions in iron bars.

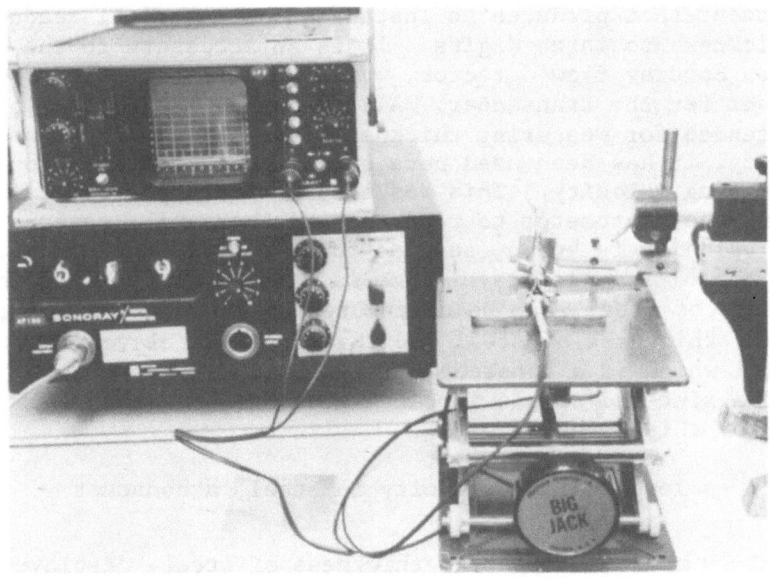

Figure 10.   Instrumentation and test setup for point by
point measurement of ultrasonic velocity in tensile bars.

ULTRASONIC TECHNIQUE FOR DETERMINING PROPERTY
VARIATIONS WITHIN A COMPONENT

In a contact ultrasonic test, the sound beam can be considered to be confined to a projection equal to the face size of the transducer. Since the area under test is determined by the size of the crystal transducer, ultrasonic velocity measurements may be confined to a localized area or critical section of a component. Therefore, if ultrasonic velocity was measured point by point along the surface of a sintered material, property variability could readily be detected.

Instrumentation and test setup utilized for point by point measurement of ultrasonic velocity of powder metal tensile bars is shown in Figure 10. A 1/4 inch 2.25 MHz dual element transducer is accurately positioned over the test bar and adjusted vertically so that a moderate pressure is exerted on the face of the bar. A small amount of glycerin is spread over the face of the bar to provide coupling. The transducer is moved along the axis of the bar and measurements are made every tenth of an inch with the Branson Digital Micrometer. The Digital Micrometer is a thickness measuring instrument that produces an instantaneous numerical readout of thickness to three digits. It is an accessory to the Branson Sonoray flaw detector, which acts as the pulser and receiver for the transducer. Although the instrumentation is intended for measuring thickness in materials of known velocity, it has been used here on sintered materials for determining velocity. This was accomplished by first calibrating the micrometer to read the thickness of a material whose velocity is known, such as fully consolidated steel. Since the ultrasonic travel time is linearly related to the thickness of steel, can be calculated by knowing the measured apparent thickness of steel and the longitudinal velocity in steel which is a constant. The ultrasonic travel time (t) in a sintered sample is calculated from equation (8). Where for this method

V_L = longitudinal velocity in steel, a constant

T = measured apparent thickness of steel, displayed on Digital Micrometer

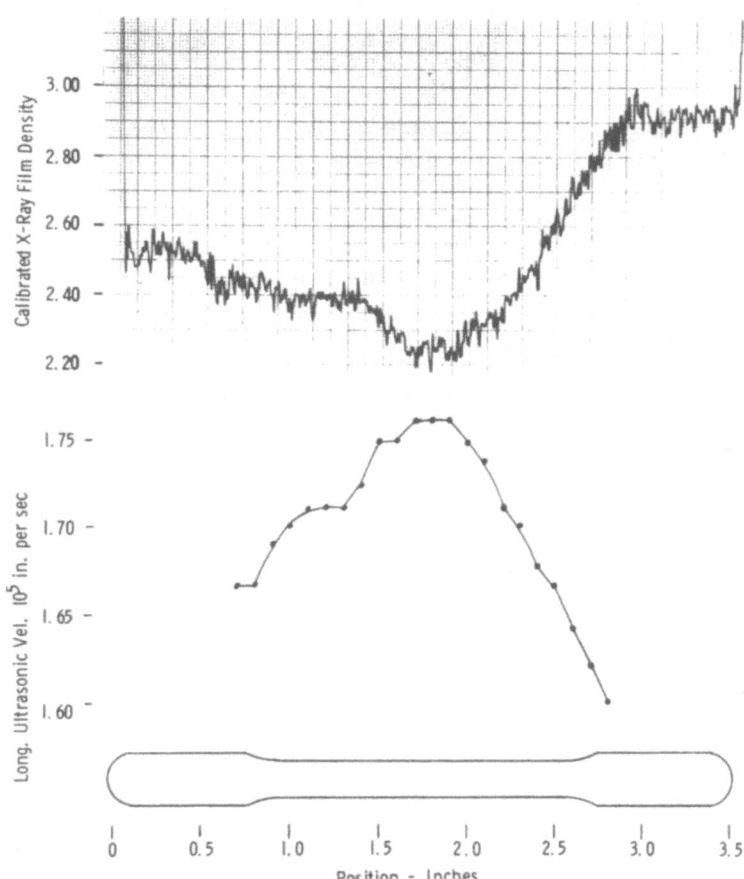

Figure 11. Results of traversing sintered tensile bar: top - variation in X-ray density; bottom - variation in ultrasonic velocity.

This time is then used with the actual specimen thickness, measured by a mechanical micrometer, to calculate velocity. The results of traversing the length of a powder metal tensile bar are shown in Figure 11. The ultrasonic velocity is highest in the center of the gage length and decreases as the bar widens. A tensile bar with this strength distribution would be expected to fracture outside its 1" gage length.

The upper curve in Figure 11 shows the results of scanning an X-ray film of the test bar with a Joyce-Loebl automatic recording microdensitometer. This instrument plots relative optical densities of photographic or X-ray film. A comparison of the two curves shows an excellent correlation of film density and ultrasonic velocity. Ultrasonic velocity traverses, therefore, it appears, can be used to detect and locate property variations in sintered powder metals. The practical application of such a method for evaluating sintered powder metal components could be enhanced by the development of a scanning system similar to those presently used to depict variations in ultrasonic attenuation. An ultrasonic velocity scanner will include a means for automatic X-Y scanning by immersion techniques, instrumentation for instantaneous measurement of transit time, and a recorder to produce a permanent three dimensional record of position and velocity during scanning.

REFERENCES

1. Coble, R. L. and Kingery, W. D., "Effect of Porosity on Physical Properties of Sintered Alumina", J. Am. Ceramic Soc., Vol. 39, No. 1, 377-385 (1956).

2. Ali, M. A., Knapp, W. J., and Kurtz, P., "Strength of Sintered Specimens Containing Hollow Glass Microspheres", Am. Ceram. Soc. Bulletin, Vol. 46, No. 3, 275-277, (1967).

3. Knudsen, F. P., "Dependence of Mechanical Strength of Brittle Polycrystalline Specimens on Porosity and Grain Size", J. Am. Ceram. Soc., Vol. 42, No. 8, 377-387, (1959).

4. Goetzel, C. G., "Treatise on Powder Metallurgy", Vol. II, Interscience Publishers, Inc., New York, New York, Chapters 33 and 34 (1950).

5. McGonnagle, W. J., "Nondestructive Testing", McGraw-Hill Book Company, Inc., New York, New York (1961), pp. 279-301.

6. McMaster, R. C., "Nondestructive Testing Handbook", Vol. II, The Ronald Press Company, New York, Chapter 51 (1959).

7. Walter, G. H., "Correlation of Structure Characteristics and Resonant Frequency Measurements with the Engineering Properties of Gray Iron", SAE, Bulletin No. MT-35, May 1965.

8. Lockyer, G. E., "Investigation of Nondestructive Methods for the Evaluation of Graphite Materials", Tech. Rpt. No. AFML-TR-65-113, Wright Patterson Air Force Base, Ohio, June 1965.

9. Anderson, O. L., and Soga, N., "Elastic Constants of Small Sintered Ceramic Specimens", Tech. Rpt. No. AFML-TR-65-202, Wright-Patterson Air Force Base, Ohio, September 1965.

10. Spinner, S. and Tefft, W. E., "A Method for Determining
     Mechanical Resonance Frequencies and for Calculating
     Elastic Moduli from These Frequencies", Proc. ASTM 61,
     1221-1238 (1961).

11. Spinner, S., Knudsen, F. P., and Stone, L., "Elastic
     Constant - Porosity Relations for Polycrystalline
     Thoria", J. Research NBS, Vol. 67C, No. 1, 39-46,
     (Jan. - Mar. 1963).

12. Piatasik, R. S. and Hasselman, D. P. H., "Effect of
     Open and Closed Pores on Young's Modulus of Poly-
     crystalline Ceramics", J. Amer. Ceram. Soc., Vol. 47,
     No. 1, 50-51, 1964.

13. Knudsen, F. P., "Effect of Porosity on Young's Modulus
     of Alumina", J. Am. Ceram. Soc., Vol. 45, No. 2, 94-95,
     1962.

14. Hashin, Z. and Rosen, B. W., "The Elastic Moduli of
     Fiber-Reinforced Materials", J. App. Mech., Vol. 31,
     No. 2, 223-232, 1964.

15. Marlowe, M. O. and Wilder, D. R., "Sensitive Resonant
     Frequency Technique for the Study of Sintering Kinetics",
     J. Am. Ceram. Soc., Vol. 50, No. 3, 145-149 (1967).

16. Truel, R., Elbaum, C., and Chick, B. B., "Ultrasonic
     Methods in Solid State Physics", Academic Press, Inc.,
     New York, N. Y. pp. 78-151, (1969).

A REVIEW OF MERCURY POROSIMETRY

H. M. Rootare

School of Dentistry
University of Michigan
Ann Arbor, Michigan 48104

INTRODUCTION

The mercury intrusion method is commonly used to
characterize a porous material as to its pore-size distribution by means of a mercury porosimeter. It is a device
capable of generating suitably high pressures and measuring
the volume of mercury taken up by the pores. The volume
change with increase of pressure may be measured several
ways. The most common way has been direct visual observation
through a window in the pressure vessel called the "Sight
Gauge". It had an advantage over the remote type of measuring
devices, such as resistance wires, capacitive bridges, and
followers, in that it allowed direct visual observation of
the volume change and direct reading of the dilatometer.
The disadvantage of this method is that it is virtually
hopeless to attempt to build such a "Sight Gauge" for high
pressure operation, where the window would have to withstand
pressures above 15,000 PSI. Therefore, a more practical and
economical approach has been to use the remote read-out devices.
Platinum-Iridium resistance wire has been used in the dilatometer to relate the change in resistance to the change in
volume of the dilatometer. Another means of measuring the
volume change in the dilatometer is the mechanical follower,
that maintains contact with the mercury column as it moves
up the dilatometer stem under pressure, and relates the linear
distance traveled to the volume of mercury intruded. A third
method is the use of the capacitance bridge to measure the
change of capacitance between the column of mercury in the

225

dilatometer stem and an external shield around the penetrometer, and relate this directly to the volume change.

Literature review. It has been forty-seven years since Washburn,[52] in 1921, first suggested the use of mercury intrusion under pressure to determine the pore-size distribution of porous solids. The relation proposed by him is as follows:

$$P = \frac{-2 \, \gamma \, \cos \, \theta}{r} \tag{1}$$

where r is the radius of the pore being intruded by mercury of surface tension γ, under pressure P, and at the contact angle θ with the material under test. The above equation holds for any liquid in contact with a porous solid having a contact angle greater than 90°. This is the common capillary depression phenomenon.

The capillary law governing liquid mercury (Hg) penetration into a small pore may be derived easily.[1] If the liquid wets the wall of the capillary, Fig. 1, the liquid surface must be parallel with the wall, and the complete surface (meniscus), will be concave in shape. The Young and Laplace equation

$$P = \gamma \left(\frac{1}{R_1} + \frac{1}{R_2} \right) \tag{2}$$

can be used to express the pressure difference across the interface. Its sign is either ($\pm$) so that the pressure is less in the liquid than in the gas phase. The radii of curvature, where both are of the same sign, always lie on the side of the interface having the greater pressure.

If we have a small cylindrical capillary, the meniscus will be approximately hemispherical, and $R_1 = R_2$ and is equal to the radius of the capillary, r. Thus equation 2 reduces to

$$\Delta P = 2 \, \gamma / r \tag{3}$$

The height of the meniscus above a flat liquid surface is h, for which P = 0; thus the hydrostatic pressure drop in the column of liquid in the capillary is also equal to ΔP.

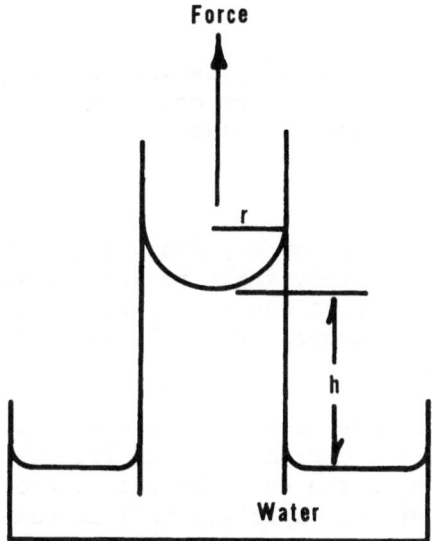

Figure 1. Capillary rise, when liquid wets the wall of the capillary, ($\theta = 0°$) concave meniscus.

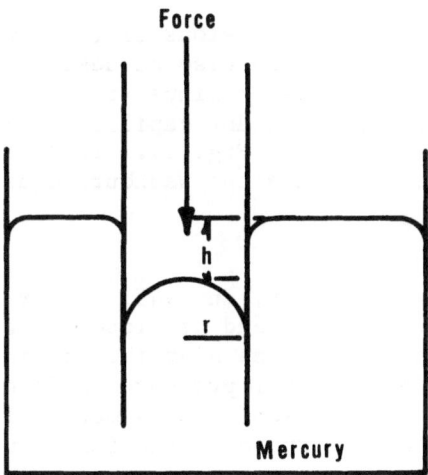

Figure 2. Capillary depression, when liquid does not wet the wall of the capillary, ($\theta = 180°$) convex meniscus.

$$\Delta P = \Delta pgh \tag{4}$$

Δp is the difference in density between the liquid and gas phase, and g is the acceleration constant due to gravity. Combining equations 3 and 4 yields

$$\Delta pgh = 2 \ \gamma/r \tag{5}$$

This equation also holds for capillary depression in a case of a liquid that completely fails to wet the walls of the capillary, and the contact angle with the wall is 180°. The meniscus will now be convex (Fig. 2), and h is now the depth of the depression.

A more general case is one somewhere in between the range of the above two extremes of complete wetting and complete nonwetting of the solid by the liquid. Ordinarily, the liquid will meet the capillary wall at some angle θ, and the relationship between the radius of curvature R and the capillary radius r is R = r/cos θ. Therefore, equation 5 becomes

$$\Delta pgh = \frac{\pm 2 \ \gamma \ \cos \ \theta}{r} \tag{6}$$

The product Δpgh has the dimensions of force per square centimeter, and can therefore also be designated as pressure P. Application of the plus or minus ($\pm$) to the equation depends on whether the resultant capillary force is up (+) or down (−), (cf. Fig. 1 and Fig. 2). Therefore, for the case of mercury we may write the Washburn equation as

$$P = \frac{-2 \ \gamma \ \cos \ \theta}{r} \tag{7}$$

Twenty-four years later, in 1945, Ritter and Drake[15,42] published experimental data and developed a method for determining the macropore size distribution in a porous solid. They applied external pressure to force the mercury (a nonwetting liquid) into some porous materials and measured the volume of mercury intruded as the function of pressure. They measured pressure-volume curves for a number of materials, for which the pore-size distributions were calculated. Developing apparatus and methods for measuring the penetration of mercury into pores down to 100 Å in radius at 10,000 PSI pressure, they introduced this method to workers interested in porous materials. By 1949 Drake[14] had increased the capacity of his equipment to reach 60,000 PSI, and thus was

able to penetrate mercury into the pores down to 18 Å in radius. He found that the surface tension concept still appeared to hold, and the maximum limit of the technique had
not been reached with mercury, whose atom is about 3.14 Å
in diameter.

The question of the contact angle value to be used in
the Washburn equation has not been established with any certainty. As n average value for many materials, Ritter and
Drake used 140°. Juhola and Wiig[30] who also were concerned
about it, drilled a hole in a coconut shell compact, and
measured the pressure required to force mercury into the 3.81
x 10^6 Å hole. In another experiment designed to test the validity of the method, mercury was forced at known pressures
into a fine, calibrated glass capillary. In both cases 140°
was found for the contact angles. Winslow and Shapiro[55]
found 130° to be a valid value for the contact angle of mercury and a standard nickel block with 70 drilled holes of
560 microns diameter. Most workers since then have used
either value for the contact angle, somewhat arbitrarily.

The reliability of measuring pore-size by mercury intrusion was proven satisfactory when compared with nitrogen
adsorption at the liquid nitrogen temperature (78°K)[3,29].
The need for defining pore structures of many important materials such as filters, adsorbents, catalysts, etc. had
become generally recognized. However, the use of the mercury intrusion method was limited to a few researchers,
where the need or the magnitude of the project justified
the effort and the expense of constructing the equipment.

Pore volume and pore size measurements have been attempted by various experimental methods. The simplest way
to determine the pore volume is by measuring the density
of the material with helium and then with mercury. The
difference between the respective specific volumes will
give directly the pore volume in cc/gram. The alternative
is to measure the pore volume directly by forcing mercury
under high pressure into the pores of the sample and reading the volume intruded from the dilatometer. Presently,
60,000 psi is the highest pressure in commercial instruments. This corresponds to a 30 Å pore diameter.

The most important direct methods are given in Table I.

TABLE I

Method	Comments
1. Density[11,21]	Difference between specific volumes of the material, measured by helium and mercury; total pore volume only, no size nor distribution.
2. Optical microscopy	Size and geometry; volume difficult; optical limit 0.5 micron in diameter.
3. Electron microscopy[20]	Size and geometry; distortion during sectioning; nonrepresentative sampling.
4. Gas adsorption[3]	Nitrogen molecule most common gas used; pore volume and size distribution for the range of pore diameters from 600 Å to 14 Å in diameter.
5. Mercury porosimetry[42]	Wider range of pore coverage; commonly attained range from 100 microns to 30 Ångströms in diameter, determined by the pressure range of the equipment.
6. Suction measuring method[12]	Water suction is measured in cm of water, or dynes/cm^2, where the entry suction depends on the pore opening size; capillary force is countered by suction.
7. X-ray small-angle scattering[43]	Extremely small (particles) or pores in the range 200 Ångströms down; existence and shape of small pores only.

The pore size distribution curves are frequently biased toward the small pore size because of the hysteresis effect caused by the bottle-necked, or the ink-well type pores. Pores of these shapes have narrow-necked openings into large volume, thus yielding erroneous information as to the true pore volume at the given size apparently measured. Meyer[38] has attempted to use probability theory to correct the mercury intrusion data to arrive at the true pore size distribution. Meyer was concerned with petroleum reservoir rock, and used physical properties of porous rocks in determining its character as an oil and gas producer. Assuming random pore distribution, he found that Poisson's distribution satisfied the derived equation. Since only the "zero" term is used, one may ignore the fact that pores are mutually exclusive in deriving Poisson's equation without introducing a serious error. A more complex problem of aggregates in a random distribution is avoided by assuming that only the pores which are isolated from all equal or larger pores will fail to fill. An example is worked out using Burdine's[6] original data to demonstrate the effect of this correction.

TABLE II[(38)]

No.	PSI	Pore Volume	ΔV	ΔV	Pore Volume Corrected Data
1.	25	0.03cc	0.03	0.14	0.14
2.	50	0.20	0.17	0.18	0.32
3.	100	0.46	0.26	0.17	0.49
4.	200	0.58	0.12	0.10	0.59
5.	500	0.67	0.09	0.08	0.67
6.	1000	0.73	0.06	0.06	0.73

It can be seen from table II that the correction has altered the distribution of the large pores considerably. However, since the probability that the smallest pores are not connected became vanishingly small, the correction at higher pressures was negligible.

Thirty-eight years after it was first proposed by Washburn[52], an instrument for routine determination of pore-size distributions by mercury intrusion was designed by Winslow and Shapiro[55]. Concurrently, in Germany, Guyer[24] developed a commercial apparatus for the automatic determination of pore-size distributions. The commercialization

of these instruments made the technique available to a large
number of workers. Prior to this time many modifications of
Ritter and Drake's[42] apparatus had been reported in the literature,[5,6,37,40,42,49,53] but compressed nitrogen was always
used to apply pressure to the mercury column. Higher pressures
were obtained by additional compression of the gas by means
of a high pressure oil pump forcing oil into the gas reservoir.[49] The change of volume in the dilatometer with increasing pressure was usually measured by using a Wheatstone-
bridge circuit. The change of resistance of the platinum-
iridium wire was measured as the mercury moved up the stem
of the dilatometer, exposing the wire with each incremental
pressure increase.

Winslow and Shapiro simplified the operation of the
porosimeter, and improved the safety of the equipment by
substituting isopropyl alcohol as the hydraulic fluid for
compressed nitrogen gas. The design of a calibrated stem
penetrometer that could be visually observed through a window
in a high pressure 'sight-gauge' simplified the measurement
of the volume data and reduced the time for analysis. The
use of hydraulic fluid and a hand pressure generator replaced
the gas cylinder and the auxiliary pump. Another feature of
this instrument was the "filling device" for initial filling
of the penetrometer with mercury, operating in the vacuum-to-
atmospheric pressure range.[8,22] The upper limit of pores
measured at reduced pressure was extended to about 100 microns,
which corresponds roughly to 2 psi above the pressure necessary to support the mercury column in the penetrometer.

Many papers have been published reporting data on a
variety of different materials: paper[27,37,54], soft woods
(pine, spruce, cedar)[47,48], textiles[7,41,51], leather[31,49],
PVC resins[19], macroreticular ion exchange resins [11,33],
carbons[4,16,17,29,30,45], coal[57], coke[8], membrane filters[28],
Florida filter clays[26], clay building bricks[53], petroleum
reservoir rocks[5,6,38,40], phosphate rocks and triple super
phosphate[9,21], refractory materials[50], porous iron[56], aluminum oxides[13], catalysts and other porous materials[13,14,
15,24,25,42,43,55].

Much work has been conducted in characterizing porous
materials by the mercury intrusion method that is proprie-
tary in nature and has not been published.

As a modification of the cylindrical capillary pore

model, Kruyer[20,32], and then Frevel and Kressley[22] derived
theoretical porosimetry curves for various packings of solid
spheres. The original model, based on the assumption of a
system of circular capillaries, does not always give a satisfactory picture when dealing with small particles and powders,
[2,20]. For porous solids prepared from nonporous powder
particles, the "inter-connected void" model developed by
Frevel and Kressley is preferred. Mercury intrusion measurements on different agglomerations of uniform microspheres
and of like-mesh particles confirm the validity of this
(Frevel and Kressley) model. The characteristic criterion
for a mechanically packed assemblage of like-mesh particles
is an abrupt threshold mercury penetration with only a slight
increase in pressure, determined by the mesh size and contact
angle. This sudden penetration of interconnected voids is
followed by the gradual filling of the toroidal voids around
the contacting particles. See Figures 3 and 4.

The treatment defines the pressure necessary to "break
through" in terms of the largest access opening to the interior
of the solid, and it relates the size of the opening to the
radius of the spheres comprising the solid rather than to
the radius of the equivalent cylindrical pore.

Calculation of "equivalent cylindrical pore radii"
from surface area data and total pore volume ($r=2V/A$), was
suggested by Emmett and DeWitt.[18] With the availability of
mercury porosimetry equipment, porous solids have often been
characterized by comparing surface area equivalent pore radii
to porosimetry equivalent pore radii[46,56]. Conversely, the
surface area values can be used to calculate an "equivalent
spherical radius" for powder particles. Frevel and Kressley's
model, comprised of uniform spheres, now allows another direct
comparison of surface area equivalent spherical radius with
the mercury porosimeter equivalent radius.

This model has recently been further refined by Mayer
and Stowe[35,36]. In order to cover a greater range of porosities, a more generalized model was developed, consisting of
uniform spheres varying in packing between the two extremes
of three-dimensional close packing and three-dimensional
simple cubic packing. This model has the same lower porosity
limit, 25.95%, as the previous model of Frevel and Kressley,
but its upper porosity limit, 47.67%, is increased from the
previous limit of 39.54%.

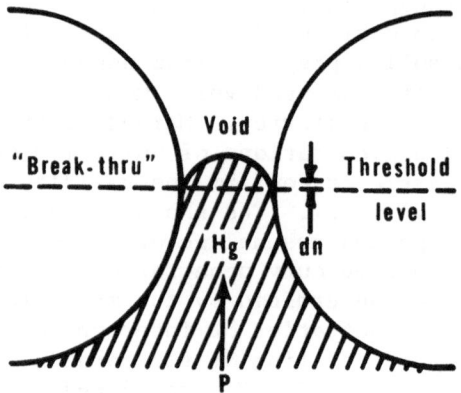

Figure 3. Mercury penetration between spheres at "break-through."

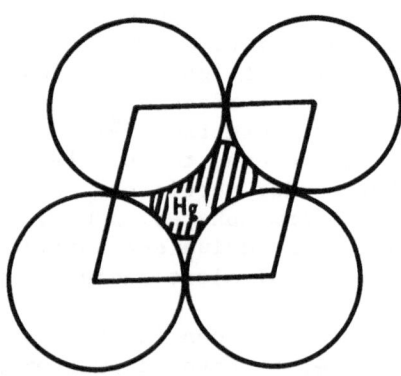

Figure 4. Access opening to unit void and toroidal voids.

The model of cylindrical pore geometry has also been utilized in calculating the value for the surface area of the sample from the total pore volume and the average pore radius obtained from the mercury porosimetry curve. Assuming n number of cylindrical pores of length 1, we can simply establish the relationship:

$$\text{Pore Volume:} \quad V = n\pi r^2 1$$

$$\text{Surface Area:} \quad A = n2\pi r 1$$

$$\text{Therefore:} \quad A = 2V/r$$

Or, it can be represented as an integral of the total internal pore surface area as:

$$A = \int_{r\,=\,0}^{r\,=\,\infty} \frac{2}{r} \left(\frac{dV}{dr}\right)\, dr$$

The comparison of pore areas obtained in this manner with the surface areas obtained by nitrogen absorption (B.E.T. Method) has been fair with simple symmetrical pore-size distributions.[56,57] However, it becomes poor when the distribution curves are skewed and more complex, as, for example, bimodal distributions.

An attempt to obtain surface area measurements from the mercury intrusion data, with no assumption of any specific pore geometry, was made by Rootare and Prenzlow.[44] The problem was approached from the point of view that work is required to force mercury into the pores, measured as PV work. This could be directly related to the total area of the sample immersed in mercury. The work required to immerse an area dA of powder (solid) surface is equal to

$$dW = \gamma_{LV} \cos \theta\ dA = -PdV \tag{8}$$

The work is supplied in the porosimeter by applying external pressure to force mercury to fill a given volume of void. Therefore, to measure the total work required to immerse the surface in mercury, it is necessary to integrate over the entire range of pressure required to fill the smallest pores present in the sample. Thus the relationship is obtained

$$A = \frac{-\int_{Vo}^{Vm} PdV}{\gamma_{LV} \cos \theta} \tag{9}$$

If the porosimetry measurements are made at room temperature, and γ_{LV} = 485 dynes/cm, and θ = 130°, the equation reduces to

$$A = \frac{+0.0221}{m} \int_{Vo}^{Vm} PdV \tag{10}$$

The surface area will be in square meters per gram of sample, if P is measured in PSIA and m is the sample weight in grams.

FACTORS CONTRIBUTING TO ERRORS IN MERCURY POROSIMETRY

1. The effect of compressibility of mercury with increasing pressure on the reading of the penetrometer stem, which has a total readable (100 graduations) displacement volume of 0.200 ml, is tabulated for two different penetrometer sample volume sizes.
        These calculations are based on the full volume of mercury with no sample in the bulb. Appropriate corrections should be made for high pressure readings, corresponding to the actual volume of mercury in the sample bulb, i.e. total volume minus the sample volume.

TABLE III

No.	Total Sample Volume, ml.	Mercury Compression in % of full scale at			
		5000 PSI	15,000 PSI	30,000 PSI	60,000 PSI
1.	1.7	1.0%	3.0%	6.0%	12.0%
2.	6.1	3.6	10.7	21.4	42.9

2. Compressibility of the solid is another factor that may cause errors in the volume measurement; however, the porosimetry technique, on the other hand, can be used to evaluate the compressibility of the solid. Zweitering and van Krevelen[57] determined, from the slope of the experimental

points above 7,500 PSI that lie on a straight line, the
compressibility of coal to be 1.3×10^{-11} cm^2/dyne. This
value compares well with an independently determined value
of 1.9×10^{-11} cm^2/dyne by the velocity of ultrasonic waves
in coal.

Compressibility of most solid materials has values
that range approximately from 10^{-11} to 10^{-12} cm^2/dyne. However, for some porous materials the compressibility may be
higher because of the presence of open or closed pores.

3. After the evacuation of the penetrometer (with or
without the sample) has been discontinued, residual air in
the penetrometer can add to the errors of intrusion volume
measurement. For example, if the 1.7 ml penetrometer has
been evacuated to 100 torr pressure, the compressibility of
the residual entrapped air at 5,000 PSI will amount to about
1% of the full scale reading of 0.200 ml.

4. Kinetic hysteresis effect, where a time factor enters
into reading of the mercury penetration before equilibrium
has been approached, may result in shifting of the distribution
curves. Sometimes considerable time is required for the mercury to flow into the pores at a given pressure. This time
factor may be reduced or eliminated by careful vibrating or
knocking of the penetrometer. The effect is very visible if
one taps on the filling device and observes the rise of mercury at a given constant pressure.

5. Volume hysteresis, in contrast to the kinetic hysteresis, is the retention of mercury by the pores of the sample
after penetration and reduction of pressure to one atmosphere.
This may be caused by the classic 'ink-bottle' type pores,
or by some other shape pores with constricted 'necks' opening
into large void volumes. The pore diameter calculated by
the Washburn equation is the diameter of the opening of the
pore and not necessarily the largest diameter of the pore;
the latter depends on the pore's shape.

6. Assuming a cylindrical pore shape model for the
analysis of mercury porosimetry data may not best represent
the actual porous material under study. Most powdered,
crystalline or spherical shaped materials will give anything
but cylindrically shaped pores. Some indication as to the
shape of the pores or the material will help in interpreting
the porosimetry data. For non-circular cross sections, the
ratio of r/2, of area to perimeter, will change because the
constant 1/2 will change for the relationship:

$$\frac{\text{Total pore volume}}{\text{Area x average radius}} = \frac{V_p}{A_p \times r} = 0.500 \text{ for cylindrical pore.}$$

Shape factors for a few simple pore models are listed in Table IV. This obviously oversimplifies the problem of the shapes of capillaries, as is evident from deBoer's[20] (p. 20 in Colston papers) discussion of fifteen shape groups of capillaries. However, it is illustrative of the problem involved in interpretation of the poresize distribution curves obtained by mercury porosimetry.

TABLE IV

Pore model, shape	Shape factor
Cylindrical	0.500
Cubic	0.304
Tetragonal	0.144
Hexagonal	0.117

7. Another variable factor resulting from the assumption of some constant value for the surface tension of mercury (γ_{Hg}) at the ambient temperature of the experiment is that the temperature coefficient of the surface tension of mercury $d\gamma/dT = 0.21$ dynes/cm°C, according to Roberts, and it agrees very well with values obtained by other workers. (See Fig. 5) However, the degree of purity of mercury has a much greater effect on the surface tension of mercury. Increase of contaminants will lower the surface tension of mercury. This may very well be the main factor in the wide spread of reported values for the surface tension of mercury in the literature. Therefore, the certainty of literature values is yet to be established. The values reported have a spread of about one hundred dynes. Kernaghan[c] (1931) reported 432.2±0.3 dynes/cm at 31°C, while Cook[b] (1929) had reported 515/dynes/cm at the same 31°C temperature.

Although Harkins' and Ewing's[a] (1920) values were reported in the International Critical Tables, 476.1 dynes/cm at 20°C, and have been widely used in mercury porosimetry calculations, a more recent and perhaps a better value appears to be about 485 dynes/cm at 25°C, reported by Roberts[g] in 1964. Table V gives some indication of the higher values obtained in the past forty years, by various workers, for the surface tension of mercury.

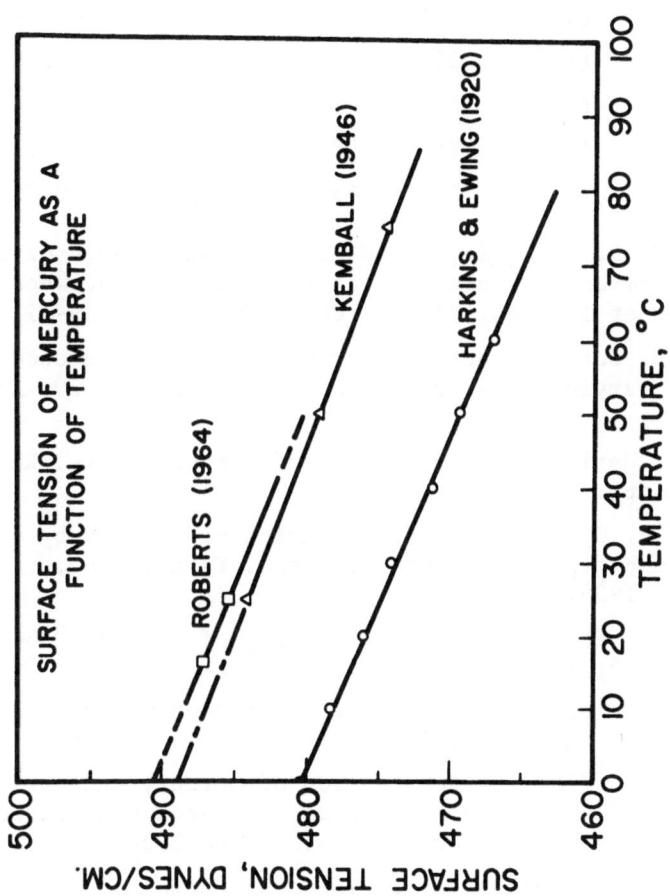

Figure 5. Dependence of surface tension of mercury as a function of temperature.

TABLE V

No.	Investi- gators		Temp., °C	Surface Tension γH_g dynes/cm	Temp. Coeff. $d\gamma/dT$, dynes/ cm, °C
1.	Owen & Dufton[h]	(1925)	17°	485 ± 1	- - - -
2.	Bosworth[d]	(1938)	20	484	0.21
3.	Young[e]	(1941)	20	485 ± 5	- - - -
4.	Kemball[f]	(1946)	25	484.2 ± 1.5	0.19
5.	Roberts[g]	(1964)	25	485.4 ± 1.2	0.21

[a]Harkins, W. D. and Ewing, W. W. The surface energy of mercury and the energy relations at the interface between mercury and other liquids.   J. Am. Chem. Soc., 42:2539-47, 1920.

[b]Cook, S. G. Surface tension of mercury in the presence of gas under varying pressures. Phys. Rev., 34:513-20, Aug. 1929.

[c]Kernaghan, Marie. Surface tension of mercury. Phys. Rev., 37:990-7, Apr. 1931.

[d]Bosworth, R. C. L. The surface tension of mercury by the maximum bubble pressure method. Trans. Faraday Soc., 34:1501-5, 1938.

[e]Young, T. F. in "Physics and Chemistry of Surfaces" by Adam, W. K., p. 185, 1941.   Oxford Univ. Press.

[f]Kemball, Charles. On the surface tension of mercury. Trans. Faraday Soc., 42:526-37, 1946.

[g]Roberts, N. K. The surface tension of mercury and the absorption of water vapor and some satu ated hydrocarbons on mercury. J. Chem. Soc., 1907-15, June 1964.

[h]Owen, E. A. and Dufton, A. F. The application of radiography to the study of capillarity. Proc. Phys. Soc., 38:204-6, 1925.

    The selection of the surface tension value for mercury determines the constant in the Washburn equation ($DP = -4\ \gamma \cos \theta$) in pore diameter calculation. For example, when the contact angle is taken as 130° for the mercury and the solid interface, the constant will vary as shown in table VI.

TABLE VI

Surface tension	473	476	480	485 dynes/cm
DP = K	176	177.5	179	181
For contact angle, θ = 130°.				

The effect on the calculated pore-opening diameter of
the choice of the above values can be seen from the tabulation
of the equivalent cylindrical pore diameters for a fixed final
absolute pressure (PSIA).

TABLE VII

Surface tension	473	476	480	485 dynes/cm
	Pore Diameters			
P = 1.8 PSIA	97.8	98.6	99.4	100.5 microns
P = 15,000	11.8	11.8	11.9	12.1 milli microns
P = 60,000	29.4	29.6	29.8	30.2 Angströms
For contact angle, θ = 130°.				

We can see from Table VII that the selection of the
surface tension value of mercury to be used in the Washburn
equation is not very critical. A deviation in the value of
surface tension of mercury in the order of 2.5% (i.e. from
473 to 485 dynes/cm) will directly result in the equivalent
change in the calculated pore diameter. Considering all
other possible sources of error, this is rather small, and
perhaps the least troublesome. One can attempt to correct
for this by measuring the surface tension of his mercury
prior to using it. And, after the run has been completed,
determine if contamination of mercury in contact with the
sample has occurred, resulting in corresponding lowering of
the surface tension.

8. The largest source of error in calculation of the
equivalent cylindrical pore diameter from the mercury pene-
gration data comes from the assumption of some constant,
average value for the contact angle θ for the mercury and
the solid interface. Since this information is not always
readily available, most workers usually assume either 130°
or 140°, depending on their preference. For comparing porous
materials of the same type, it does not really matter which
number one chooses, unless he is attempting to obtain an
exact measure of the pore openings. In this case, he should
either measure the contact angle directly or look at the pores

under the optical or the electron microscope, to establish
the relationship for the actual pore size opening to that
measured by mercury intrusion.

Table VIII demonstrates the extent of the range of
equivalent cylindrical pore-diameter values one can have
when there is a difference between the true and the assumed
average contact angle. For a wide range of materials the
contact angles may vary easily from 110° to 160°. In it
are calculated equivalent cylindrical pore diameters at three
different pressures for various possible contact angles.

TABLE VIII

Contact angle θ°	Pressures in PSIA	(γ = 485 dynes/cm)	
	1.8	15,000	60,000
110°	53.5 microns	6.42 milli	16.0 Angströms
120	78.2	9.38 microns	23.4
130	100.	12.1	30.2
140	120.	14.4	35.9
150	135.	16.2	40.6
160	147.	17.6	44.1

It can be seen from table VIII that the calculation of
the actual pore opening diameter hinges to a large extent on
the selection of the contact angle value. The agreement
between the calculated value and the actual value for the
pore diameter depends on how close to the real value the
selected average value of the contact angle happens to be.
One can miss the actual pore diameter by 50% or 100% in an
extreme case, by choosing an average value of either 130°
or 140° for the contact angle, if, for example, the real
value is either 160° or 110°.

9. Breakdown of porous materials during the pressuri-
zation of the sample to force the mercury into the pores
can be a source of error in porosity study. When mercury is
forced into the pores, the larger pores may be subject to
breakdown, especially if they have bottleneck type constricted
openings. For equal wall thickness, the smaller pores are
much stronger than the larger pores. Sponge or foam type
materials are especially subject to this type of 'squeeze'
effect because their walls are thin and of low mechanical
strength.

Summary and Discussion. Most of the work reported in
the literature concerns the application of mercury intrusion
to porosity characterizations of a variety of materials.
This method is a natural outgrowth of the Young and Laplace
capillary law. It has taken almost fifty years for the
technique to become well established in the evaluation of
materials.

The mercury intrusion method was introduced by
Washburn in 1921. Ritter and Drake were the first to publish
some data, in 1945, and to develop the method for macropore-
size distribution of porous solids. The method did not
immediately become widely used because of the highly specialized
nature and the expense of the equipment. In 1959 Winslow
and Shapiro developed the first commercialized instrument in
the United States for Mercury Porosimetry. This began the
era of voluminous data collection of pore-size distributions
on many and varied types of porous and particulate materials.

Frevel and Kressley have innovated a new "inter-connected
toroidal void model" of porosity between packed spheres to
replace the more idealized model of cylindrical shaped pores.
Mayer and Stowe have extended this model by increasing the
upper porosity limit from 39.54% to 47.67%. The new model
consists of uniform spheres with a variation in packing between the two extremes of three-dimensional close packing
and the three-dimensional simple cubic packing. They have
shown that theoretical curves which were derived from the
models of different three-dimensional packings of like-mesh
spherical particles yield a more valid model of the actual
porosities measured. This would be especially true when
applied to powders of nonporous particles, compacted pelletts
of powders, or sintered porous solids made of nonporous
particles.

A procedure also was developed by Mayer and Stowe for
calculating the "equivalent spherical particle radius" using
the porosity and the "break through" pressure obtained from
the mercury porosimetry technique.

Based on the assumption of cylindrical pore geometry,
mercury intrusion data have been used to estimate the surface
areas of powders from the total pore volume and the "equivalent mean pore radius." Frequently, this has been a good
approximation, and has saved the additional effort and time
involved in measuring surface area by the accepted B.E.T.

method of gas adsorption. However, Rootare and Prenzlow
demonstrated that surface areas can be calculated from mercury pore-size distribution curves without an assumption of
certain pore geometry. The extent to which a surface area
calculated in this manner may compare with one calculated
by the B.E.T. method depends largely on the closeness of the
assumed contact angle to the actual value. If the contact
angle is measured directly on the material, and this value
is used in the surface area calculation, much closer agreement with the B.E.T. value may be expected.

It is evident that one of the most important factors in
the application of mercury intrusion data is the value selected
for the interfacial contact angle between the mercury and the
solid. The closer we can estimate the true contact angle
value, the more accurate will be out subsequent calculations.

Contact angles are normally difficult to measure
accurately on porous, granular, or powdered materials. Powders could be compressed into tablets, and the contact angle
measured directly on the tablet surface; however, the values
obtained should be corrected for the roughness of the surface.
For contact angles greater than 90°, surface roughness will
increase the measured contact angle value in proportion to
the degree of surface roughness.

The mercury intrusion technique has been exploited
only from the standpoint of practical application to porosity
measurements. No attempt appears to have been made concerning
the use of this technique as an additional method of obtaining
contact angle data for particulate or powdered materials.

Since the capillary laws still apply when mercury under
pressure is intruded into the pores or the interstitial voids
of particles, equation No. 9 can be used to solve for cos θ.
Suppose we measure the surface area of the sample by the
B.E.T. method, and insert it into the equation. The only
other unknown parameter is the contact angle, and we can
solve for it.

$$\cos \theta = \frac{-0.01422}{A} \int_{V_o}^{V_m} P dV \tag{11}$$

This should give a much more reliable value for the contact
angle of mercury on the actual surface of the particular
solid under consideration. Tabulation of some contact angle

values obtained in this manner is given in table IX.

TABLE IX

No.	Material	Surface Areas		Cosθ	θ
		Mercury $\theta = 130°$	Nitrogen		
1.	Aluminum powder	$1.35 m^2/g$	$1.14 m^2/g$	0.7611	140°
2.	Anatase, Titanium Dioxide	15.1	10.3	0.9423	160°
3.	Carbon Black (Spheron 6)	107.8	110.0	0.6299	129°
4.	Carbon, Sterling FT (2700°F)	15.7	12.3	0.8202	145°
5.	Copper powder	0.34	0.49	0.4461	116°
6.	Glass, Alkali Borosilicate	11.0	7.9	0.8948	153°
7.	Iron powder	0.20	0.30	0.4287	115°
8.	Tungsten powder	0.11	0.10	0.7071	135°
9.	Tungsten carbide powder	0.111	0.14	0.5097	121°
10.	Zinc powder	0.34	0.32	0.6826	133°

Once the contact angle has been calculated for the material, this value can then be used in the subsequent pore-size distribution determinations of a material of the same kind, but of different porosity or particle size. Thus more exact pore-size diameter values can be calculated from the absolute pressures, and these values should closely approach the actual opening diameters of the pores. The surface areas calculated using the new contact angle value should have one-to-one correspondence to the surface areas measured by the B.E.T. method. If it is of importance to obtain information about the surface area of the sample, calculating it from the pore-size distribution curve saves considerable time compared to the additional effort involved in the B.E.T. surface area measurement.

CONCLUSIONS

1. Mercury intrusion method for pore-size distribution measurement is a well established method in the study of porous solids.

2. The ability of the method to yield accurate information about the size of the pores and their shape depends on several factors and assumptions:

   a) compressibilities of the solid, mercury, and residual air remaining in the sample space,
   b) breakdown of porous materials under pressure,
   c) kinetic hysteresis or the time effect, and the volume hysteresis or the pore shape effect,
   d) assumption of cylindrical pore model,
   e) assumption of constant or a given value for the surface tension of mercury, and finally
   f) the assumption of some constant value for the contact angle between mercury and the solid.

3. The mercury porosimetry technique can also be used for the evaluation of particle size, surface area, and contact angle.

REFERENCES

1. Adamson, A. W., Physical Chemistry of Surfaces, 2nd ed.,
    New York, Interscience, 1967, 697 p. (The Mercury
    Porosimeter, p. 546-9.)

2. Barrer, R. M., McKenzie, N., and Reay, J. S. S., Capil-
    lary Condensation in Single Pores, J. Colloid Sci.,
    11:479-95, Jan. 1956.

3. Barrett, E. P., Joyner, L. G., and Halenda, P. P., The
    Determination of Pore Volume and Area Distributions in
    Porous Substances. I. Computations from Nitrogen Iso-
    therms, J. Am. Chem. Soc., 73:373-80, 1951.

4. Bond, R. L., ed., Porous Carbon Solids, New York,
    Academic Press, 1968, 444 p.

5. Bucker, H. P., Jr., Felsenthal, M. , and Conley, F. R.,
    A Simplified Pore-Size Distribution Apparatus, J.
    Petroleum Technology, AIME, 8:65-6, Apr. 1956.

6. Burdine, N. T., Gournay, L. S., and Reichertz, P. P.,
    Pore-Size Distribution of Petroleum Reservoir Rocks,
    Petroleum Transactions, AIME, 189: 195-204, 1950.

7. Burleigh, E. G., Wakeham, Helmut, Honald, Edith, and
    Skau, E. L., Pore-Size Distribution in Textiles, Tex.
    Res. J., 19: 547-55, Sept. 1949.

8. Cameron, A., and Stacy, W. O., Apparatus and Technique,
    A Low Pressure Mercury Porosity Meter, Che. Ind.
    (London), 222-23, Feb. 27, 1960.

9. Caro, J. H., and Freeman, H. P., Pore Structure of
    Phosphate Rock and Triple Superphosphate, J. Agr. Food
    Chem., 9: 182-86, May-June 1961.

10. Cartan, F. A., and Curtis, G. J., Apparatus for the
    Determination of Particle Density of Porous Solids,
    Anal. Chem., 35: 423-24, Mar. 1963.

11. Cassidy, H. G., and Kun, K. A., Oxidation-Reduction
    Polymers: Redox Polymers, New York, Wiley-Inter-
    science, 1965, 307 p. (4.5.4 Pore Structure, p. 155-67.)

12. Ceaglske, N. H., and Hougen, O. A., The Drying of
    Granular Solids, Trans. Am. Inst. Chem. Engrs., 33:
    283-314, 1937.

13. Cochran, C. N., and Cosgrove, L. A., Pore-Size Distri-
    bution of Porous Aluminum Oxides by Mercury Porosimeter
    and n-Butane Sorption, J. Phys. Chem., 61: 1417-19, Oct.
    1957.

14. Drake, L. C., Pore-Size Distribution in Porous
    Materials. Application of High Pressure Mercury
    Porosimeter to Crackling Catalysts, Ind. Eng. Chem.,
    41: 780-85, Apr. 1949.

15. Drake, L. E., and Ritter, H. L., Macropore-Size Distri-
    bution in Some Typical Porous Substances, Ind. Eng.
    Chem. Anal. Ed., 17: 787-91, Dec. 1945.

16. Dubinin, M. M., Porous Structure and Adsorption Proper-
    ties of Active Carbons, p. 51-120. Walker, P. L., Jr.,
    ed. Chemistry and Physics of Carbon, A Series of Ad-
    vances, Vol. 2, New York, Marcel Dekker, 1966, 384 p.

17. Dubinin, M. M., Vishnyakova, M. M., Zhukovskaya, E. G.,
    Leont'ev, E. A. Luk'yanovich, V. M., and Sarakhov, A.
    I., Investigation of the Porous Structure of Solids by
    Sorption Methods. V. The Structure of Intermediate
    Pores and Macropores of Activated Carbons, Russ. J.
    Phys. Chem. (Engl. Edn.) 34: 959-64, Sept. 1960.

18. Emmett, P. H., and DeWitt, T. W., The Low Temperature
    Adsorption of Nitrogen, Oxygen, Argon, Hydrogen, n-
    Butane and Carbon Dioxide on Porous Glass and Partially
    Dehydrated Chabazite, J. Am. Chem. Soc., 65: 1253-62,
    1943.

19. Engle, D. L., The Internal Pore Structure of Polyvinyl
    Chloride Resins, Union Carbide Report, Research and
    Development Dept., Chemicals Div., Union Carbide Corp.,
    South Charleston, West Virginia, Oct. 22, 1963.

20. Everett, D. H., and Stone, F. S., eds. The Structure
    and Properties of Porous Materials, London, Butter-
    worth, 1958, 382 p. (Vol. 10 of the Colston Papers.)

21. Freeman, H. P., Care, J. H., and Heinly, N., Effect of
     Calcination on the Character of Phosphate Rock, Agric.
     and Food Chem., 12: 479-86, Nov. - Dec. 1964.

22. Frevel, L. K., and Kressley, L. J., Modifications in
     Mercury Porosimetry, Anal. Chem., 35: 1492-1501,
     Sept. 1963.

23. Gregg, S. J., and Sing, K. S. W., Adsorption, Surface
     Area and Porosity, New York, Academic Press, 1967,
     355 P. (Estimation of Pore-Size Distribution by the
     Mercury Porosimeter, p. 182-4).

24. von Guyer, A., Jr., Bohlen, B., and Guyer, A., Uber die
     Bestimmung von Porengrossen, Helv. Chim. Acta, 42:
     2103-10, 1959.

25. Hashinguchi, Yukio, Urano, Yokichi, and Iwasaka, Masaji,
     Prevention of the Explosion of Dissolved Acetylene. II.
     Pore Distribution in Porous Materials Used as Containers
     for Dissolved Acetylene, Koatsu Gasu, 4: No. 4, 197-200,
     1967, Japan.

26. Henderson, L. M., Ridgway, C. M., and Ross, W. B., Some
     Variables in Filter Clays, Refiner, and Natural Gaso-
     line Manufacturer, 19: No. 6, 69-74, June 1940.

27. Herman, D. F., and Dunlap, I. R., Polyethylene En-
     capsulated Cellulose. A New Papermaking Fiber, Tappi,
     48: 418-23, July 1965.

28. Honald, Edith, and Skau, E. L., Application of Mercury
     Intrusion Method for Determination of Pore-Size Distri-
     bution to Membrane Filters, Science, 120: 805-6, Nov.
     12, 1954.

29. Joyner, L. G., Barrett, E. P., and Skold, R., The De-
     termination of Pore Volume and Area Distributions in
     Porous Substances. II. Comparison Between Nitrogen
     Isotherm and Mercury Porosimeter Methods, J. Am. Chem.
     Soc., 73: 3155-8, July 1951.

30. Juhola, A. J., and Wiig, E. O., Pore Structure in Acti-
     vated Charcoal. II. Determination of Macro Pore-Size
     Distribution, J. Amer. Chem. Soc., 71: 2078-80, June
     1949.

31. Kanagy, J. R., Macro-Pores in Leather as Determined with a Mercury Porosimeter, J. Am. Leather Chemists' Assoc., 58: 524-50, Sept. 1963.

32. Kruyer, S., The Penetration of Mercury and Capillary Condensation in Packed Spheres, Trans. Faraday Soc., 54: 1758-67, 1958.

33. Kun, K. A., and Kunin, Robert, Pore Structure of Some Macroreticular Ion Exchange Resins, Polymer Letters, 2: 587-91, 1964.

34. Loisy, R., A Method of Studying Porosity, Bull. Soc. Chim., 8: 569, July-Aug. 1941. (French)

35. Mayer, R. P., and Stowe, R. A., Mercury Porosimetry. Breakthrough Pressure for Penetration Between Packed Spheres, J. Colloid Sci., 20: 893-911, 1965.

36. Mayer, R. P., and Stowe, R. A., Mercury Porosimetry. Filling of Toroidal Void Volume Following Breakthrough Between Packed Spheres, J. Phys. Chem., 70: 3867-73, Dec. 1966.

37. McKnight, T. S., Marchassault, R. H., and Mason, S. G., The Distribution of Pore-Sizes in Wood-Pulp Fibres and Paper, Pulp and Paper Magazine of Canada, 59: No. 2, 81-8, Feb. 1958.

38. Meyer, H. I., Pore Distribution in Porous Media, J. Appl. Phys., 24: 510-12, May 1953.

39. Neher, M. B., Extended Range Hudraulic Mercury Porosimeter, Anal. Chem., 33: 1132-36, July 1961.

40. Purcell, W. R., Capillary Pressures--Their Measurement Using Mercury and the Calculation of Permeability Therefrom, J. Petrol. Technology, 1: 39-48, Feb. 1949.

41. Quynn, R. G., Internal Volume in Fibers, Textile Res. J., 33: 21-34, Jan. 1963.

42. Ritter, H. L., and Drake, L. C., Pore-Size Distribution in Porous Materials; Pressure Porosimeter and Determinations of Complete Macropore-Size Distribution, Ind. Eng. Chem. Anal. Ed., 17: 782-6, Dec. 1945.

43. Ritter, H. L., and Erich, L. C., Pore-Size Distribution
    in Porous Materials, Anal. Chem., 20: 665-70, July 1948.

44. Rootare, H. M., and Prenzlow, C. F., Surface Areas from
    Mercury Porosimeter Measurements, J. Phys. Chem.,
    71: 2734-36, July 1967.

45. Sarakhov, A. I., Some Comments on the Accuracy of the
    Method of Mercury Porometry, Russ, J. Phys. Chem.
    (Engl. Edn.) 37: 242-43, Feb. 1963.

46. Shull, C. G., The Determination of Pore-Size Distribution
    from Gas Adsorption Data, J. Am. Chem. Soc., 70: 1405-
    10, Apr. 1948.

47. Stayton, C. L., and Hart, C. A., Determining Pore-Size
    Distribution in Softwoods with Mercury Porosimeter,
    Forest Prod. J., 15: 435-40, Oct. 1965.

48. Stone, J. E., Scallon, A. M., and Aberson, G. M. A.,
    The Wall Density of Native Cellulose Fibres, Pulp and
    Paper Magazine of Canada, 67: No. 5, T 263-68, May,
    1966.

49. Stromberg, R. R., Pore-Size Distribution in Collagen
    and Leather by the Porosimeter Method, J. Res. N. B. S.,
    54: 73-81, Feb. 1955.

50. Ulmer, G. C., and Smothers, W. J., Application of Mer-
    cury Porosimetry to Refractory Materials, Am. Cer. Soc.
    Bul., 46: 649-52, July 1967.

51. Wakeham, Helmut, and Spicer, Nancy, Pore-Size Distri-
    bution in Textiles. A Study of Windproofing and Water-
    Resistant Cotton Fabrics, Tex. Res. J., 19: 703-10,
    Nov. 1949.

52. Washburn, E. W., Note on a Method of Determining the
    Distribution of Pore Sizes in a Porous Material, Proc.
    Nat. Acad. Sci., 7: 115-16, 1921.

53. Watson, A., May J. O., and Butterworth, B., Studies of
    Pore-Size Distribution. I. Apparatus and Preliminary
    Results, Trans. Brit. Ceramic Soc., 56: 37-50, Feb. 1957.

54. White, R. E., and Marceau, W. E., The Capillary Be-
    havior of Paper, Tappi, 45: 279-84, Apr. 1962.

55. Winslow, N. M., and Shapiro, J. J., An Instrument for
    the Measurement of Pore-Size Distribution by Mercury
    Penetration, ASTM Bull., TP 49: 39-44, Feb. 1959.

56. Zwietering, P., and Koks, H. L. T., Pore-Size Distribu-
    tion of Porous Iron, Nature, 173: 683-4, Apr. 1954.

57. Zwietering, P., and van Krevelen, D. W., Chemical
    Structure and Properties of Coal. IV. Pore Structure.
    Fuel, 33: 331-7, 1954.

INDEX